Inquiry in Action
Interpreting Scientific Papers

for

Campbell Biology
Ninth Edition

Reece • Urry • Cain • Wasserman • Minorsky • Jackson

Part A: Scientific Papers and Questions for Inquiry Figures
in *Campbell Biology*, Ninth Edition, and
Campbell/Reece *Biology*, Eighth Edition

Ruth Buskirk
University of Texas at Austin

Part B: Reading Primary Literature:
A Practical Guide to Evaluating Research Articles in Biology

Christopher M. Gillen
Kenyon College

Benjamin Cummings

Boston   Columbus   Indianapolis   New York   San Francisco   Upper Saddle River
Amsterdam   Cape Town   Dubai   London   Madrid   Milan   Munich   Paris   Montréal   Toronto
Delhi   Mexico City   São Paulo   Sydney   Hong Kong   Seoul   Singapore   Taipei   Tokyo

Vice President/Editor-in-Chief: Beth Wilbur
Acquisitions Editor: Josh Frost
Senior Editorial Manager: Ginnie Simione Jutson
Senior Supplements Project Editor: Susan Berge
Managing Editor, Production: Michael Early
Production Project Manager: Jane Brundage
Manufacturing Buyer: Michael Penne
Executive Marketing Manager: Lauren Harp
Production Management and Composition: S4Carlisle Publishing Services
Cover Design Production: Seventeenth Street Studios
Text and Cover Printer: Courier, Kendallville

Cover Photo Credit: "Succulent I" ©2005 Amy Lamb, www.amylamb.com

ISBN-13: 978-0-321-68336-6
ISBN-10: 0-321-68336-6

Benjamin Cummings
is an imprint of

Credits

Article 1, page 2, Inquiry Figure 1.27: Reprinted by permission from Macmillan Publishers, Ltd: D. W. Pfennig, W. R. Harcombe, and K. S. Pfennig, Frequency-dependent Batesian mimicry, *Nature* 410:323 (2001).

Article 2, page 6, Inquiry Figure 2.2: Reprinted by permission from Macmillan Publishers, Ltd: M. E. Frederickson, M. J. Greene, and D. M. Gordon, "Devil's gardens" bedevilled by ants, *Nature* 437:495–496 (2005).

Article 3, page 11, Inquiry Figure 16.11: M. Meselson and F. W. Stahl, The replication of DNA in *Escherichia coli, Proceedings of the National Academy of Sciences USA* 44:671–682 (1958).

Article 4, page 26, Inquiry Figure 23.16: Reprinted with permission from AAAS: A.M. Welch et al., Call duration as an indicator of genetic quality in male gray tree frogs, *Science* 280:1928–1930 (1998).

Article 5, page 32, Inquiry Figure 37.14: K. A. Stinson et al., Invasive plant suppresses the growth of native tree seedlings by disrupting belowground mutualisms, *PLoS Biol* (*Public Library of Science: Biology*) 4(5): e140 (2006).

Article 6, page 40, Inquiry Figure 41.4: Reprinted with permission from Elsevier: R. W. Smithells et al., Possible prevention of neural-tube defects by periconceptional vitamin supplementation, *Lancet* 315: 339–340 (1980).

Article 7, page 45, Inquiry Figure 56.13: Reprinted with permission from AAAS: R.L. Westemeier et al., Tracking the long-term decline and recovery of an isolated population. *Science* 282:1695–1698 (1998).

Article 8, page 52, Inquiry Figure 9: Reprinted by permission of Macmillan Ltd: H. Itoh, et al., Mechanically driven ATP synthesis by F_1-ATPase, *Nature* 427:465–468 (2004).

Article 9, page 59, Inquiry Figure 30: R. D. Sargent, Floral symmetry affects speciation rates in angiosperms, *Proceedings of the Royal Society B: Biological Sciences*, 271:603–608 (2004).

Contents

PART A

SCIENTIFIC PAPERS AND QUESTIONS FOR INQUIRY FIGURES IN *CAMPBELL BIOLOGY,* NINTH EDITION (INQUIRY FIGURES 1.27, 2.2, 16.11, 23.16, 37.14, 41.4, AND 56.13), AND CAMPBELL/REECE *BIOLOGY,* EIGHTH EDITION (INQUIRY FIGURES 9 AND 30)

RUTH BUSKIRK

ARTICLE 1

Inquiry Figure 1.27: *Does the Presence of Venomous Coral Snakes Affect Predation Rates on Their Mimics, Kingsnakes?*

Introduction—The Article and Phenomenon Under Study

Many poisonous animals have warning coloration that signals to potential predators they are dangerous. Sometimes a harmless species, with warning coloration that mimics the dangerous species, benefits when predators confuse them with the harmful species. This phenomenon is called *Batesian mimicry*. Batesian mimicry should only be effective if predators have experience with the dangerous species. In order to test this mimicry hypothesis in nature, investigators designed field experiments with coral snakes and their mimics. This research, from the following paper, is the subject of Inquiry Figure 1.27 in *Campbell Biology*, Ninth Edition:

D. W. Pfennig, W. R. Harcombe, and K. S. Pfennig, Frequency-dependent Batesian mimicry, *Nature* 410: 323 (2001).

Read the complete article beginning on the next page and then answer the questions following the article.

▼ Figure 1.27

INQUIRY

Does the presence of venomous coral snakes affect predation rates on their mimics, kingsnakes?

EXPERIMENT David Pfennig and his colleagues made artificial snakes to test a prediction of the mimicry hypothesis: that kingsnakes benefit from mimicking the warning coloration of venomous coral snakes only in regions where coral snakes are present. The researchers placed equal numbers of artificial kingsnakes (experimental group) and brown artificial snakes (control group) at 14 field sites, half in the area the two snakes cohabit and half in the area where coral snakes are absent. The researchers recovered the artificial snakes after four weeks and tabulated predation data based on teeth and claw marks on the snakes.

RESULTS In field sites where coral snakes are absent, most attacks were on artificial kingsnakes. Where coral snakes were present, most attacks were on brown artificial snakes.

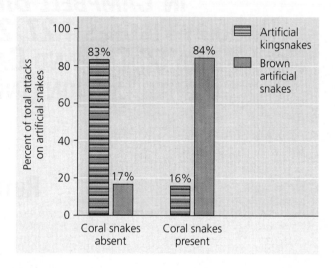

CONCLUSION The field experiments support the mimicry hypothesis by not falsifying the prediction, which was that mimicking coral snakes is effective only in areas where coral snakes are present. The experiments also tested an alternative hypothesis: that predators generally avoid all snakes with brightly colored rings. That hypothesis was falsified by the data showing that in areas without coral snakes, the ringed coloration failed to repel predators. (The fake kingsnakes may have been attacked more often in those areas because their bright pattern made them easier to spot than the brown fakes.)

SOURCE D. W. Pfennig, W. R. Harcombe, and K. S. Pfennig, Frequency-dependent Batesian mimicry, *Nature* 410:323 (2001).

INQUIRY IN ACTION Read and analyze the original paper in *Inquiry in Action: Interpreting Scientific Papers.*

(MB) See the related Experimental Inquiry Tutorial in MasteringBiology.

WHAT IF? What experimental results would you predict if predators throughout the Carolinas avoided all snakes with brightly colored ring patterns?

brief communications

Frequency-dependent Batesian mimicry

Predators avoid look-alikes of venomous snakes only when the real thing is around.

Batesian mimicry holds that palatable species look like dangerous species because both are then protected from predation[1-5]. But this protection should break down where the dangerous model is absent, when predators would not be under selection to recognize the model or any other species resembling it as dangerous[2,4,5]. Here we provide experimental evidence to support this critical prediction of Batesian mimicry by demonstrating that predators avoid harmless look-alikes of venomous coral snakes only in areas that are inhabited by these deadly snakes.

Many coral snakes and non-venomous kingsnakes possess red, yellow (or white), and black ringed markings[6], which predators avoid[7], though often without prior experience[8]. To determine whether this avoidance depends on the model's presence in the vicinity, we constructed snake replicas[7] (1.5 cm × 18 cm cylinders of pre-coloured, non-toxic plasticine threaded onto an S-shaped wire) with a tricolour ringed pattern, a striped pattern with identical colours and proportions as the ringed replicas, or a plain brown pattern.

Ringed replicas conformed to the local mimic: scarlet kingsnakes (*Lampropeltis triangulum elapsoides*), which resemble eastern coral snakes (*Micrurus fulvius*)[9], or sonoran mountain kingsnakes (*L. pyromelana*), which resemble western coral snakes (*Micruroides euryxanthus*)[10]; striped and brown replicas served as controls. We arranged three different replicas (triplets) 2 m apart in natural habitat (each was used once only). At each site, 10 triplets were placed 75 m apart in a line. After collection, a person without knowledge of the replica's location scored attacks by noting any impressions corresponding to a predator[7].

We tested whether predators avoid *L. t. elapsoides* only in areas inhabited by *Micrurus* by placing 10 triplets at eight sympatric sites (sites with mimic and model) and eight allopatric sites (sites with only the mimic) in North and South Carolina, USA (480 replicas; allopatric sites were more than 80 km outside *Micrurus*'s range[9,11]; sites were 16–420 km apart). After 4 weeks, 25 (5.2%) replicas had been attacked by carnivores. The mean (± s.e.m.) proportion of ringed replicas attacked was significantly greater in allopatry (0.654 ± 0.107) than in sympatry (0.083 ± 0.116; $P = 0.009$, 2-tailed Wilcoxon two-group test).

We next investigated whether predators avoid *L. pyromelana* only in sympatry with *Micruroides* by placing 10 triplets at 24 sites (720 replicas) along an elevational gradient

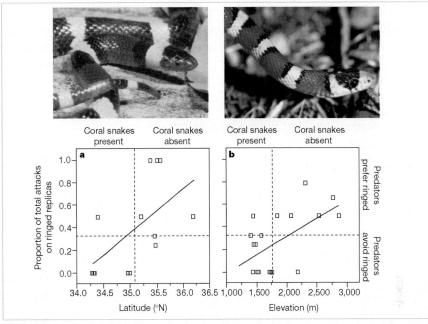

Coral snakes present — Coral snakes absent

Coral snakes present — Coral snakes absent

a Proportion of total attacks on ringed replicas (y-axis: 0.0–1.0); Latitude (°N) (x-axis: 34.0–36.5)

b Elevation (m) (x-axis: 1,000–3,000); Predators prefer ringed / Predators avoid ringed

Figure 1 Frequency-dependent mimicry. The proportion of carnivore attacks on ringed replicas of scarlet kingsnakes (top left; a mimic of eastern coral snakes) and sonoran mountain kingsnakes (top right; a mimic of western coral snakes) increased with **a**, latitude ($y = -13.314 + 0.391x$, $P < 0.035$, $R^2 = 0.345$) and **b**, elevation ($y = -0.329 + 0.00032x$, $P < 0.014$, $R^2 = 0.310$). Horizontal dashed line: proportion of attacks on ringed replicas expected under randomness. Vertical dashed line: maximum latitude and elevation for coral snakes in North Carolina and Arizona, respectively.

(1,204–2,866 m) near Portal, Arizona (*Micruroides* only occur at altitudes below 1,770 m (ref. 10); there were 14 sympatric and 10 allopatric sites 3–100 km apart). After 2 weeks, 49 (6.8%) replicas had been attacked by carnivores.

The mean proportion of ringed replicas attacked was significantly greater in allopatry (0.496 ± 0.078) than in sympatry (0.138 ± 0.060; $P = 0.006$). Moreover, in sympatry, the proportion of ringed replicas attacked (0.138) was significantly less than randomness (0.33; $P = 0.010$, 2-tailed Wilcoxon signed-rank test). By contrast, attacks were random in allopatry ($P = 0.188$). Thus, predators avoid coral snake mimics only in sympatry with the model.

Coral snakes become increasingly rare with increasing latitude (Spearman $\rho = -0.57$, $P = 0.014$)[11] and elevation ($\rho = -0.77$, $P = 0.026$; our unpublished results). Consequently, selection to avoid ringed patterns should weaken with increasing latitude and elevation. As expected, the proportion of ringed replicas attacked increased gradually with latitude and elevation (Fig. 1), suggesting that selection to avoid ringed patterns is indeed sensitive to the abundance of coral snakes.

Our results do not fully resolve why mimetic patterns occur where models are absent[6,9–11]. Possibly selection for mimicry

in sympatry, coupled with gene flow between sympatric and allopatric populations[12], maintains mimetic patterns in both regions. Nevertheless, our results verify the critical prediction of Batesian mimicry and demonstrate that the benefits of mimicry depend on abundance of the model.

David W. Pfennig*, William R. Harcombe*, Karin S. Pfennig†

**Department of Biology, University of North Carolina, Chapel Hill,*
North Carolina 27599-3280, USA
e-mail: dpfennig@email.unc.edu

†*Section of Integrative Biology, University of Texas, Austin, Texas 78712-1064, USA*

1. Bates, H. W. *Trans. Linn. Soc. Lond.* **23**, 495–566 (1862).
2. Wallace, A. R. *Contributions to the Theory of Natural Selection* (Macmillan, London, 1870).
3. Fisher, R. A. *The Genetical Theory of Natural Selection. A Complete Variorum Edition* (Oxford University Press, New York, 1999).
4. Waldbauer, G. P. *Evol. Biol.* **22**, 227–259 (1988).
5. Maynard Smith, J. *Evolutionary Genetics* (Oxford University Press, Oxford, 1998).
6. Greene, H. W. & McDiarmid, R. Y. *Science* **213**, 1207–1212 (1981).
7. Brodie, E. D. *Evolution* **47**, 227–235 (1993).
8. Smith, S. M. *Science* **187**, 759–760 (1975).
9. Conant, R. & Collins, J. T. *A Field Guide to Reptiles and Amphibians of Eastern and Central North America* (Houghton Mifflin, Boston, MA, 1998).
10. Stebbins, R. C. *Western Reptiles and Amphibians* (Houghton Mifflin, Boston, MA, 1985).
11. Palmer, W. M. & Braswell, A. L. *Reptiles of North Carolina* (University of North Carolina Press, Chapel Hill, NC, 1995).
12. King, R. B. & Lawson, R. *BioScience* **47**, 279–286 (1997).

Guiding Questions for Reading This Article

A. About the Article

1. Give the name of the journal and the year in which this article was published.

2. What are the last names of the three authors? At what university was the work done?

3. Specialized vocabulary: Write a brief definition of each term.

   Defined in the article:

   Batesian mimicry

   Not defined in the article:

   sympatric

   allopatric

4. What type of organism is being studied? Give genus and species names, as well as common names, for two of the study species.

5. This study is designed to test what prediction of Batesian mimicry?

B. About the Study

6. From what materials did the investigators make the experimental models?

7. At each study site, investigators placed how many snake models of what three color patterns?

8. How many sympatric sites and how many allopatric sites were tested in North Carolina and South Carolina? How many sympatric sites and how many allopatric sites were tested in Arizona?

9. In Figure 1a, what is the *x*-axis? What is the *y*-axis? Which is the dependent variable? In Figure 1b, what is the *x*-axis and what is the *y*-axis?

10. What are the patterns of coral snake presence and absence by latitude and by elevation?

11. Hypothesis: Predators avoid Batesian mimics only in areas that are inhabited by the dangerous model.

    (a) Prediction (a) under this hypothesis: The proportion of total attacks on ringed replicas at *latitudes* where coral snakes are present will be _____ (higher? lower? no different?) than at latitudes where coral snakes do not occur.

    (b) Prediction (b) under this hypothesis: The proportion of total attacks on ringed replicas at *elevations* where coral snakes are present will be _____ (higher? lower? no different?) than at elevations where coral snakes do not occur.

12. Null hypothesis: There is no relationship between predator avoidance of Batesian mimics and presence of the dangerous model.

    (a) Prediction (a) under this null hypothesis: The proportion of total attacks on ringed replicas at *latitudes* where coral snakes are present will be _____ (higher? lower? no different?) than at latitudes where coral snakes do not occur.

(b) Prediction (b) under this null hypothesis: The proportion of total attacks on ringed replicas at *elevations* where coral snakes are present will be _____ (higher? lower? no different?) than at elevations where coral snakes do not occur.

13. In Figure 1a, look at the proportion of total attacks on ringed replicas placed at different latitudes in North Carolina and South Carolina. In which areas are the attack rates on ringed replicas higher: in areas where coral snakes are present or in areas where coral snakes are absent?

14. Do the results in Figure 1a (#13) agree with prediction (a) under the hypothesis?

15. Do the results in Figure 1a (#13) agree with prediction (a) under the null hypothesis?

16. The results in Figure 1a lead us to do which of these? (A) reject the hypothesis; (B) reject the null hypothesis.

17. In Figure 1b, look at the proportion of total attacks on ringed replicas placed at different elevations. In which areas are the attack rates on ringed replicas higher: in areas where coral snakes are present or in areas where coral snakes are absent?

18. Do the results in Figure 1b (#17) agree with prediction (b) under the hypothesis?

19. Do the results in Figure 1b (#17) agree with prediction (b) under the null hypothesis?

20. The results in Figure 1b lead us to do which one of these? (A) Reject the hypothesis; (B) reject the null hypothesis.

21. Is this an *observational study*, in which quantitative, observational data are taken but no experimental manipulation is made, or is this an *experimental study*, in which researchers make manipulations by which the effects of different variables are tested, one at a time?

22. Is this a *field study*, with data collected on organisms in their natural habitat, or is this a *lab study*, in which animals are studied under controlled conditions in the laboratory?

C. General Conclusions and Extensions of the Work

23. This system to measure predation on model snakes allows us to test specific predictions about Batesian mimicry. It is possible that other factors, besides the advantages of mimicry, explain the results observed. Perhaps it is simply the combination of bright red, yellow, and black colors on the snake replicas—not the ringed pattern itself—that explains the difference in attack rates. How could investigators test that possibility?

24. What if a particular milk snake subspecies is a poor mimic of the coral snake? Make a prediction: If this test is repeated in a geographic area where the milk snakes do not resemble coral snakes at all, would more ringed replicas be attacked?

25. Imagine that you were a member of this research team and involved in these experiments. What could be a possible follow-up test that extends this work? Briefly state another experiment or measurement you would do within this research system.

Inquiry Figure 2.2: *What Creates "Devil's Gardens" in the Rain Forest?*

Introduction—The Article and Phenomenon Under Study

Tropical rain forests have many species of trees, with such a great diversity that it is rare to have several individuals of the same species next to each other. A notable exception is in the Amazonian "devil's gardens," which consist of large stands of a single species of tree, *Duroia hirsuta*. What creates these "devil's gardens"? Inquiry Figure 2.2 in *Campbell Biology*, Ninth Edition, presents experimental data testing possible chemical mechanisms from the following paper:

M. E. Frederickson, M. J. Greene, and D. M. Gordon, "Devil's gardens" bedevilled by ants, *Nature* 437:495–496 (2005).

Read the complete article beginning on the next page and then answer the questions following the article.

▼ Figure 2.2

INQUIRY

What creates "devil's gardens" in the rain forest?

EXPERIMENT Working under Deborah Gordon and with Michael Greene, graduate student Megan Frederickson sought the cause of "devil's gardens," stands of a single species of tree, *Duroia hirsuta*. One hypothesis was that ants living in these trees, *Myrmelachista schumanni*, produce a poisonous chemical that kills trees of other species; another was that the *Duroia* trees themselves kill competing trees, perhaps by means of a chemical.

To test these hypotheses, Frederickson did field experiments in Peru. Two saplings of a local nonhost tree species, *Cedrela odorata*, were planted inside each of ten devil's gardens. At the base of one sapling, a sticky insect barrier was applied; the other was unprotected. Two more *Cedrela* saplings, with and without barriers, were planted about 50 meters outside each garden.

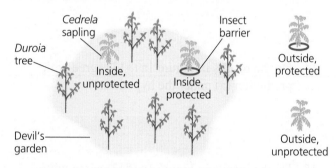

The researchers observed ant activity on the *Cedrela* leaves and measured areas of dead leaf tissue after one day. They also chemically analyzed contents of the ants' poison glands.

RESULTS The ants made injections from the tips of their abdomens into leaves of unprotected saplings in their gardens (see photo). Within one day, these leaves developed dead areas (see graph). The protected saplings were uninjured, as were the saplings planted outside the gardens. Formic acid was the only chemical detected in the poison glands of the ants.

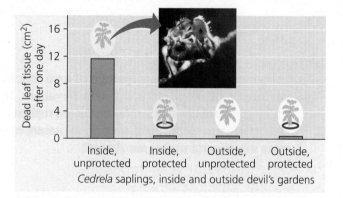

Cedrela saplings, inside and outside devil's gardens

CONCLUSION Ants of the species *Myrmelachista schumanni* kill nonhost trees by injecting the leaves with formic acid, thus creating hospitable habitats (devil's gardens) for the ant colony.

SOURCE M. E. Frederickson, M. J. Greene, and D. M. Gordon, "Devil's gardens" bedevilled by ants, *Nature* 437:495–496 (2005).

INQUIRY IN ACTION Read and analyze the original paper in *Inquiry in Action: Interpreting Scientific Papers.*

WHAT IF? What would be the results if the unprotected saplings' inability to grow in the devil's gardens was caused by a chemical released by the *Duroia* trees rather than by the ants?

Vol 437|22 September 2005

nature

BRIEF COMMUNICATIONS

'Devil's gardens' bedevilled by ants

An ant species uses herbicidal weaponry to secure its own niche in the Amazonian rainforest.

'Devil's gardens' are large stands of trees in the Amazonian rainforest that consist almost entirely of a single species, *Duroia hirsuta*[1-5], and, according to local legend, are cultivated by an evil forest spirit. Here we show that the ant *Myrmelachista schumanni*, which nests in *D. hirsuta* stems, creates devil's gardens by poisoning all plants except its host plants with formic acid. By killing these other plants, *M. schumanni* provides its colonies with abundant nest sites — a long-lasting benefit as colonies can live for 800 years.

M. schumanni lives in the hollow, swollen stems (domatia) of *D. hirsuta*, the tree species that dominates devil's gardens (Fig. 1a). Previous studies of the mutualism between *D. hirsuta* and *M. schumanni* indicated that devil's gardens result from allelopathy, which is the local inhibition of plant growth by another plant, by *D. hirsuta*[2-5]. However, studies of a different ant–plant mutualism — between an unidentified species of *Myrmelachista* and the ant-plants *Tococa guianensis* and *Clidemia heterophylla* — indicated that *Myrmelachista* may create stands comprising only its host plants by using herbicide[6,7].

We did an ant-exclusion experiment to determine whether the selective killing of plants inside devil's gardens is due to the activity of *M. schumanni* workers or to allelopathy by *D. hirsuta*. We planted saplings of a common Amazonian tree, the cedar *Cedrela odorata*, inside and outside devil's gardens, and either excluded or did not exclude ants from the saplings (for methods, see supplementary information).

We found that the *M. schumanni* workers

Figure 1 | The ant *M. schumanni* creates devil's gardens by killing all plants other than its host tree, *D. hirsuta*. a, A devil's garden, or mono-specific stand of *D. hirsuta*, in the foreground contrasts with the species-rich rainforest in the background. **b,** A worker *M. schumanni* ant attacking a plant: the ant bites a small hole in the leaf tissue, inserts the tip of its abdomen into the hole and releases formic acid. **c,** Leaves develop necrosis along primary veins within hours of the attack.

promptly attacked the saplings in devil's gardens from which ants had not been excluded, injecting a poison into their leaves (Fig. 1b), which developed necrosis within 24 hours (Fig. 1c). Most of the leaflets on these saplings were lost within five days, and the proportion lost was significantly higher than on saplings from which ants were excluded (Fig. 2). We also found that ant-free *C. odorata* inside devil's gardens fared as well as *C. odorata* planted outside devil's gardens. These results show that devil's gardens are produced by *M. schumanni* workers, rather than by *D. hirsuta* allelopathy.

In a second experiment, we investigated whether *M. schumanni* attacks only plants that are not its host plants and whether the ant uses domatia to recognize its host. We planted *C. odorata* saplings with and without artificial domatia and *D. hirsuta* saplings with and without domatia in devil's gardens. After 24 h, there was significant leaf necrosis on all *C. odorata* plants (mean area on plants with artificial domatia: 39.7 cm^2; s.e., 26.4–55.6 cm^2; on plants without artificial domatia: 14.2 cm^2; s.e., 9.2–20.3 cm^2), whereas there was no leaf necrosis at all on any *D. hirsuta* plants, irrespective of the presence of domatia (analysis of variance, $F_{3,20} = 57.03$, $P \ll 0.0001$). We conclude that *M. schumanni* attacks only non-host plants such as *C. odorata* and that it does not rely on the presence of domatia to discriminate between its hosts and other plant species.

Chemical analysis revealed that the poison glands of *M. schumanni* contain formic acid (0.43 ± 0.12 μl per worker); no other compounds were detected. Treatment of leaves with formic acid induced leaf necrosis on all the plants we tested. (For details, see supplementary information.) Many formicine ants produce formic acid: to our knowledge, this is the first record of an ant using formic acid as a herbicide — although it is known to have bactericidal and fungicidal properties[8].

Devil's gardens covered 4.5% of our study plot and grew by $0.7 \pm 0.3\%$ per year. Using this growth rate, we estimate that the largest

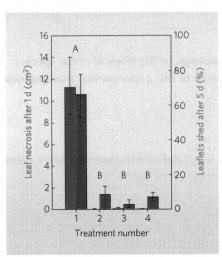

Figure 2 | *M. schumanni* ants, and not allelopathy, create devil's gardens. Saplings of the non-host plant *C. odorata* were subjected to different treatments: 1, planted inside a devil's garden, ants not excluded; 2, planted inside a devil's garden, ants excluded; 3, planted outside devil's gardens, ants not excluded; and 4, planted outside devil's gardens, ants excluded. Only saplings exposed to ants inside devil's gardens developed significant necrosis within one day (average ± s.e.; blue bars) and shed a significant percentage of their leaflets within five days (average ± s.e.; red bars). Multivariate analysis of variance results: Pillai trace, 0.88, $F_{6,72} = 9.41$, $P \ll 0.0001$. Analysis of variance (ANOVA) results (necrosis): $F_{3,36} = 52.78$, $P \ll 0.0001$. ANOVA results (leaflets shed): $F_{3,36} = 17.19$, $P \ll 0.0001$. Bars marked A are significantly different ($P < 0.001$) by Tukey post-hoc tests from bars marked B.

BRIEF COMMUNICATIONS   NATURE|Vol 437|22 September 2005

devil's garden in our plot, with 351 plants, is 807 years old (95% confidence interval, 446–4,234 years old). A devil's garden is tended by a single *M. schumanni* colony (our unpublished results) comprising as many as 3 million workers and 15,000 queens; the presence of many queens undoubtedly contributes to colony longevity.

The cultivation of devil's gardens by *M. schumanni* is an example of niche construction[9]. By killing plants of other species, the ant promotes the growth and establishment of *D. hirsuta*, thereby gaining more nest sites. A devil's garden begins when an *M. schumanni* queen colonizes a single *D. hirsuta* tree: over time, *D. hirsuta* saplings become estab-

lished within the vegetation-free area created by the ants, and the ant colony expands to occupy them. The devilry of *M. schumanni* today provides homes for ants in the future.

Megan E. Frederickson*, Michael J. Greene†, Deborah M. Gordon*
*Department of Biological Sciences, Stanford University, Stanford, California 94305-5020, USA
e-mail: meganf@stanford.edu
†Department of Biology, University of Colorado at Denver and Health Sciences Center, Denver, Colorado 80217-3364, USA

1. Davidson, D. W. & McKey, D. *Trends Ecol. Evol.* **8**, 326–332 (1993).
2. Campbell, D. G, Richardson, P. M. & Rosas Jr, A. *Biochem. Syst. Ecol.* **17**, 403–407 (1989).
3. Page, J. E., Madriñan, S. & Towers, G. H. N. *Experientia* **50**, 840–842 (1994).
4. Aquino, R., De Tommasi, N., Tapia, M., Lauro, M.R. & Rastrelli, L. J. *Nat. Prod.* **62**, 560–562 (1999).
5. Pfannes, K. R. & Baier, A. C. *Rev. Biol. Trop.* **50**, 293–301 (2002).
6. Morawetz, W., Henzl, M. & Wallnöfer, B. *Biodivers. Conserv.* **1**, 19–33 (1992).
7. Renner, S. S. & Ricklefs, R. E. *Biotropica* **30**, 324–327 (1998).
8. Revis, H. C. & Waller, D. A. *Auk* **121**, 1262-1268 (2004).
9. Odling-Smee, F. J., Laland, K. N. & Feldman, M. W. *Niche Construction: The Neglected Process in Evolution* (Princeton Univ. Press, Princeton, 2003).

Supplementary information accompanies this communication on *Nature*'s website.
Competing financial interests: declared none.
doi:10.1038/437495a

Guiding Questions for Reading This Article

A. About the Article

1. Give the name of the journal and the year in which this article was published.

2. What are the last names of the three authors? What are their universities?

3. Specialized vocabulary: Write a brief definition of each term.

   Defined in the article:

   allelopathy

   domatia

   Not defined in the article:

   formic acid

   herbicide

   host plant

   mutualism

   necrosis

   niche

4. Write out the full genus and species name of the ant in this study. Why is this scientific name written in italics in the article?

5. What is the abbreviated genus and species name for the tree in this study?

B. About the Study

6. Why are these regions of the Amazonian forest called "devil's gardens"?

7. What are two alternate hypotheses to explain why these single-species patches of trees occur in an otherwise species-rich tropical forest?

8. How were saplings of the tropical cedar species *C. odorata* used in this experimental manipulation?

9. In what two ways did investigators measure damage to the *C. odorata* cedar saplings?

10. One testable hypothesis is that the formation of "devil's gardens" involves ants, independent of allelopathy. The null hypothesis states that ants have no effect. Consider hypothesis 1 here, and fill in this chart with your predictions about leaf damage to cedar saplings, yes or no.

Hypothesis 1: *M. schumanni* ant defense keeps other plants out of the garden by increasing leaf damage to cedar saplings. Null hypothesis: There is no relationship between presence of ants and amount of leaf damage to cedar saplings.

Predictions Under Hypothesis 1	Will leaf damage occur if ants are excluded?	Will leaf loss occur if ants are not excluded?
Cedar saplings planted inside the garden (ants present)	A	B
Cedar saplings planted outside the garden (ants not present)	C	D

11. An alternate testable hypothesis is that *Duroia* allelopathy accounts for the formation of "devil's gardens," independent of ant defense. Consider hypothesis 2 here, and fill in the chart with your predictions about leaf damage to cedar saplings, yes or no.

Hypothesis 2: *D. hirsuta* tree allelopathy keeps other plants out of the garden by increasing leaf damage to cedar saplings. Null hypothesis: There is no relationship between presence of *D. hirsuta* trees and amount of leaf damage to cedar saplings.

Predictions Under Hypothesis 2	Will leaf loss occur if ants are excluded?	Will leaf loss occur if ants are not excluded?
Cedar saplings planted inside the garden (ants present)	A	B
Cedar saplings planted outside the garden (ants not present)	C	D

12. The four cells (A–D) of the table represent four treatments of the tropical cedar saplings. Why did the investigators use *more than one* sapling in each of the four treatment groups?

13. Look at the results of Figure 2 in the article (p. 495). Which of the four treatment groups of tropical cedar *C. odorata* (nonhost plant) suffered the greatest leaf necrosis after one day? Which treatment group had the greatest percentage of leaf shedding after one week?

14. Compare these results to your predictions in question 10. Do these results agree with your predictions under hypothesis 1? Do these results allow you to reject that null hypothesis?

15. Compare these results to your predictions in question 11. Do these results agree with your predictions under hypothesis 2? Do these results allow you to reject that null hypothesis?

16. Although the experimental results in Figure 2 allow you to reject hypothesis 2, they do not "prove" hypothesis 1 and do not explain how the ants cause leaf loss. How did the authors test whether the leaf damage was caused by formic acid?

17. Is this an *observational study*, in which quantitative, observational data are taken but no experimental manipulation is made? Or, is this an *experimental study*, in which researchers make manipulations by which the effects of different variables are tested, one at a time?

18. Is this a *field study*, with data collected on organisms in their natural habitat, or is this a *lab study*, in which plants are studied under controlled conditions in the laboratory or greenhouse?

20. Imagine that you were a member of this research team and involved in these experiments. What could be a possible follow-up test that extends this work? Briefly state another experiment or measurement you would do within this research system.

C. General Conclusions and Extensions of the Work

19. Explain how the behavior of *M. schumanni* ants lead to the formation of "devil's gardens."

ARTICLE 3

Inquiry Figure 16.11: *Does DNA Replication Follow the Conservative, Semiconservative, or Dispersive Model?*

Introduction—The Article and Phenomenon Under Study

When Watson and Crick published their double-helix model of DNA structure in 1953, they noted that the complementary pairing of nitrogenous bases suggested, in theory, a possible mechanism for replication. However, in practice, it was difficult to test their idea. In innovative and careful research with *E. coli* bacteria, investigators Meselson and Stahl used a novel way to label DNA and were able to confirm the mechanism of DNA replication. This research, from the following paper, is the subject of Inquiry Figure 16.11 in *Campbell Biology*, Ninth Edition:

M. Meselson and F. W. Stahl, The replication of DNA in *Escherichia coli, Proceedings of the National Academy of Sciences USA* 44:671–682 (1958).

Read the complete article beginning on the next page and then answer the questions following the article.

▼ Figure 16.11 **INQUIRY**

Does DNA replication follow the conservative, semiconservative, or dispersive model?

EXPERIMENT At the California Institute of Technology, Matthew Meselson and Franklin Stahl cultured *E. coli* for several generations in a medium containing nucleotide precursors labeled with a heavy isotope of nitrogen, ^{15}N. They then transferred the bacteria to a medium with only ^{14}N, a lighter isotope. A sample was taken after DNA replicated once; another sample was taken after DNA replicated again. They extracted DNA from the bacteria in the samples and then centrifuged each DNA sample to separate DNA of different densities.

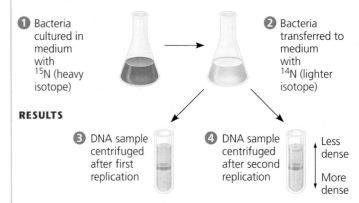

1 Bacteria cultured in medium with ^{15}N (heavy isotope)

2 Bacteria transferred to medium with ^{14}N (lighter isotope)

RESULTS

3 DNA sample centrifuged after first replication

4 DNA sample centrifuged after second replication

Less dense

More dense

CONCLUSION Meselson and Stahl compared their results to those predicted by each of the three models in Figure 16.10, as shown below. The first replication in the ^{14}N medium produced a band of hybrid (^{15}N-^{14}N) DNA. This result eliminated the conservative model. The second replication produced both light and hybrid DNA, a result that refuted the dispersive model and supported the semiconservative model. They therefore concluded that DNA replication is semiconservative.

Predictions:	First replication	Second replication
Conservative model		
Semiconservative model		
Dispersive model		

SOURCE M. Meselson and F. W. Stahl, The replication of DNA in *Escherichia coli, Proceedings of the National Academy of Sciences USA* 44:671–682 (1958).

INQUIRY IN ACTION Read and analyze the original paper in *Inquiry in Action: Interpreting Scientific Papers.*

(MB) See the related Experimental Inquiry Tutorial in MasteringBiology.

WHAT IF? If Meselson and Stahl had first grown the cells in ^{14}N-containing medium and then moved them into ^{15}N-containing medium before taking samples, what would have been the result?

Vol. 44, 1958   *BIOLOGY: MESELSON AND STAHL*   671

THE REPLICATION OF DNA IN ESCHERICHIA COLI*

By Matthew Meselson and Franklin W. Stahl

GATES AND CRELLIN LABORATORIES OF CHEMISTRY,† AND NORMAN W. CHURCH LABORATORY OF CHEMICAL BIOLOGY, CALIFORNIA INSTITUTE OF TECHNOLOGY, PASADENA, CALIFORNIA

Communicated by Max Delbrück, May 14, 1958

Introduction.—Studies of bacterial transformation and bacteriaphage infection[1-5] strongly indicate that deoxyribonucleic acid (DNA) can carry and transmit hereditary information and can direct its own replication.   Hypotheses for the mechanism of DNA replication differ in the predictions they make concerning the distribution among progeny molecules of atoms derived from parental molecules.[6]

Radioisotopic labels have been employed in experiments bearing on the distribution of parental atoms among progeny molecules in several organisms.[6-9]   We anticipated that a label which imparts to the DNA molecule an increased density might permit an analysis of this distribution by sedimentation techniques.   To this end, a method was developed for the detection of small density differences among

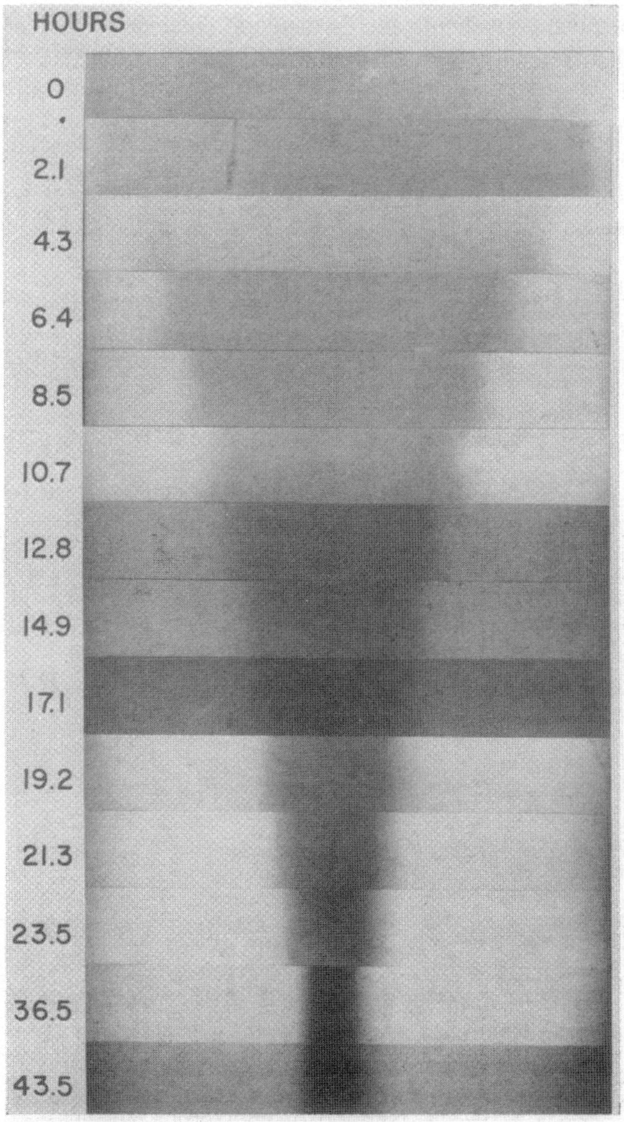

Fig. 1.—Ultraviolet absorption photographs showing successive stages in the banding of DNA from *E. coli*. An aliquot of bacterial lysate containing approximately 10^8 lysed cells was centrifuged at 31,410 rpm in a CsCl solution as described in the text. Distance from the axis of rotation increases toward the right. The number beside each photograph gives the time elapsed after reaching 31,410 rpm.

macromolecules.[10] By use of this method, we have observed the distribution of the heavy nitrogen isotope N^{15} among molecules of DNA following the transfer of a uniformly N^{15}-labeled, exponentially growing bacterial population to a growth medium containing the ordinary nitrogen isotope N^{14}.

Density-Gradient Centrifugation.—A small amount of DNA in a concentrated solution of cesium chloride is centrifuged until equilibrium is closely approached.

Vol. 44, 1958   *BIOLOGY: MESELSON AND STAHL*   673

The opposing processes of sedimentation and diffusion have then produced a stable concentration gradient of the cesium chloride. The concentration and pressure gradients result in a continuous increase of density along the direction of centrifugal force. The macromolecules of DNA present in this density gradient are driven by the centrifugal field into the region where the solution density is equal to their own buoyant density.[11] This concentrating tendency is opposed by diffusion, with the result that at equilibrium a single species of DNA is distributed over a band whose width is inversely related to the molecular weight of that species (Fig. 1).

If several different density species of DNA are present, each will form a band at the position where the density of the CsCl solution is equal to the buoyant density of that species. In this way DNA labeled with heavy nitrogen (N^{15}) may be

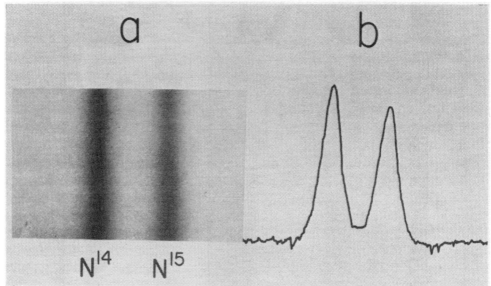

Fig. 2—*a:* The resolution of N^{14} DNA from N^{15} DNA by density-gradient centrifugation. A mixture of N^{14} and N^{15} bacterial lysates, each containing about 10^8 lysed cells, was centrifuged in CsCl solution as described in the text. The photograph was taken after 24 hours of centrifugation at 44,770 rpm. *b:* A microdensitometer tracing showing the DNA distribution in the region of the two bands of Fig. 2a. The separation between the peaks corresponds to a difference in buoyant density of 0.014 gm. cm.$^{-3}$

resolved from unlabeled DNA. Figure 2 shows the two bands formed as a result of centrifuging a mixture of approximately equal amounts of N^{14} and N^{15} *Escherichia coli* DNA.

In this paper reference will be made to the apparent molecular weight of DNA samples determined by means of density-gradient centrifugation. A discussion has been given[10] of the considerations upon which such determinations are based, as well as of several possible sources of error.[12]

Experimental.—*Escherichia coli* B was grown at 36° C. with aeration in a glucose salts medium containing ammonium chloride as the sole nitrogen source.[13] The growth of the bacterial population was followed by microscopic cell counts and by colony assays (Fig. 3).

Bacteria uniformly labeled with N^{15} were prepared by growing washed cells for

BIOLOGY: MESELSON AND STAHL   PROC. N. A. S.

14 generations (to a titer of 2×10^8/ml) in medium containing 100 μg/ml of $N^{15}H_4Cl$ of 96.5 per cent isotopic purity.   An abrupt change to N^{14} medium was then accomplished by adding to the growing culture a tenfold excess of $N^{14}H_4Cl$, along with ribosides of adenine and uracil in experiment 1 and ribosides of adenine, guanine, uracil, and cytosine in experiment 2, to give a concentration of 10 μg/ml of each riboside.   During subsequent growth the bacterial titer was kept between

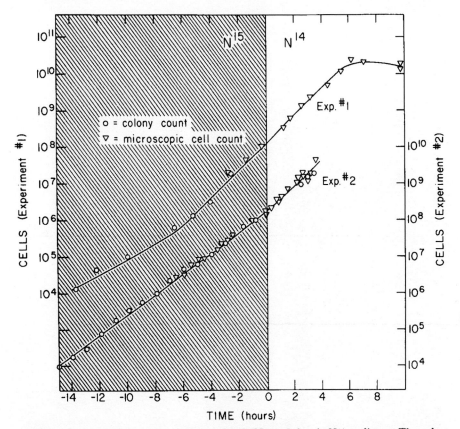

FIG. 3.—Growth of bacterial populations first in N^{15} and then in N^{14} medium.   The values on the ordinates give the actual titers of the cultures up to the time of addition of N^{14}. Thereafter, during the period when samples were being withdrawn for density-gradient centrifugation, the actual titer was kept between 1 and 2×10^8 by additions of fresh medium.   The values on the ordinates during this later period have been corrected for the withdrawals and additions.   During the period of sampling for density-gradient centrifugation, the generation time was 0.81 hours in Experiment 1 and 0.85 hours in Experiment 2.

1 and 2×10^8/ml by appropriate additions of fresh N^{14} medium containing ribosides.

Samples containing about 4×10^9 bacteria were withdrawn from the culture just before the addition of N^{14} and afterward at intervals for several generations. Each sample was immediately chilled and centrifuged in the cold for 5 minutes at $1,800 \times g$.   After resuspension in 0.40 ml. of a cold solution 0.01 M in NaCl and 0.01 M in ethylenediaminetetra-acetate (EDTA) at pH 6, the cells were lysed by the addition of 0.10 ml. of 15 per cent sodium dodecyl sulfate and stored in the cold.

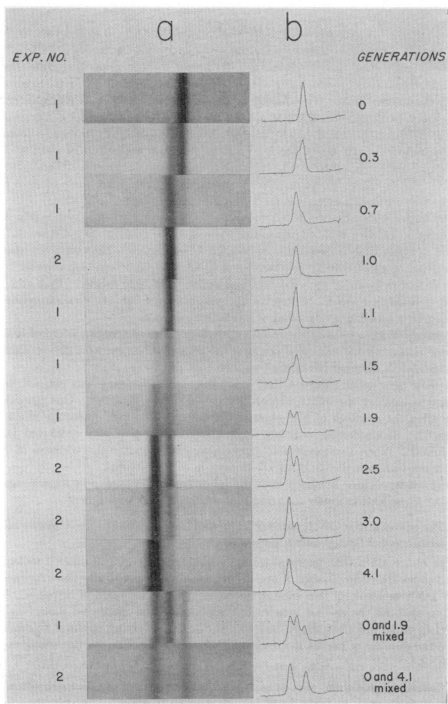

Fig. 4—*a:* Ultraviolet absorption photographs showing DNA bands resulting from density-gradient centrifugation of lysates of bacteria sampled at various times after the addition of an excess of N^{14} substrates to a growing N^{15}-labeled culture. Each photograph was taken after 20 hours of centrifugation at 44,770 rpm under the conditions described in the text. The density of the CsCl solution increases to the right. Regions of equal density occupy the same horizontal position on each photograph. The time of sampling is measured from the time of the addition of N^{14} in units of the generation time. The generation times for Experiments 1 and 2 were estimated from the measurements of bacterial growth presented in Fig. 3. *b:* Microdensitometer tracings of the DNA bands shown in the adjacent photographs. The microdensitometer pen displacement above the base line is directly proportional to the concentration of DNA. The degree of labeling of a species of DNA corresponds to the relative position of its band between the bands of fully labeled and unlabeled DNA shown in the lowermost frame, which serves as a density reference. A test of the conclusion that the DNA in the band of intermediate density is just half-labeled is provided by the frame showing the mixture of generations 0 and 1.9. When allowance is made for the relative amounts of DNA in the three peaks, the peak of intermediate density is found to be centered at 50 ± 2 per cent of the distance between the N^{14} and N^{15} peaks.

BIOLOGY: MESELSON AND STAHL   Proc. N. A. S.

For density-gradient centrifugation, 0.010 ml. of the dodecyl sulfate lysate was added to 0.70 ml. of CsCl solution buffered at pH 8.5 with 0.01 M tris(hydroxymethyl)aminomethane. The density of the resulting solution was 1.71 gm. cm.$^{-3}$ This was centrifuged at $140,000 \times g$. (44,770 rpm) in a Spinco model E ultracentrifuge at 25° for 20 hours, at which time the DNA had essentially attained sedimentation equilibrium. Bands of DNA were then found in the region of density 1.71 gm. cm.$^{-3}$, well isolated from all other macromolecular components of the bacterial lysate. Ultraviolet absorption photographs taken during the course of each centrifugation were scanned with a recording microdensitometer (Fig. 4).

The buoyant density of a DNA molecule may be expected to vary directly with the fraction of N^{15} label it contains. The density gradient is constant in the region between fully labeled and unlabeled DNA bands. Therefore, the degree of labeling of a partially labeled species of DNA may be determined directly from the relative position of its band between the band of fully labeled DNA and the band of unlabeled DNA. The error in this procedure for the determination of the degree of labeling is estimated to be about 2 per cent.

Results.—Figure 4 shows the results of density-gradient centrifugation of lysates of bacteria sampled at various times after the addition of an excess of N^{14}-containing substrates to a growing N^{15}-labeled culture.

It may be seen in Figure 4 that, until one generation time has elapsed, half-labeled molecules accumulate, while fully labeled DNA is depleted. One generation time after the addition of N^{14}, these half-labeled or "hybrid" molecules alone are observed. Subsequently, only half-labeled DNA and completely unlabeled DNA are found. When two generation times have elapsed after the addition of N^{14}, half-labeled and unlabeled DNA are present in equal amounts.

Discussion.—These results permit the following conclusions to be drawn regarding DNA replication under the conditions of the present experiment.

1. *The nitrogen of a DNA molecule is divided equally between two subunits which remain intact through many generations.*

The observation that parental nitrogen is found only in half-labeled molecules at all times after the passage of one generation time demonstrates the existence in each DNA molecule of two subunits containing equal amounts of nitrogen. The finding that at the second generation half-labeled and unlabeled molecules are found in equal amounts shows that the number of surviving parental subunits is twice the number of parent molecules initially present. That is, the subunits are conserved.

2. *Following replication, each daughter molecule has received one parental subunit.*

The finding that all DNA molecules are half-labeled one generation time after the addition of N^{14} shows that each daughter molecule receives one parental subunit.[14] If the parental subunits had segregated in any other way among the daughter molecules, there would have been found at the first generation some fully labeled and some unlabeled DNA molecules, representing those daughters which received two or no parental subunits, respectively.

3. *The replicative act results in a molecular doubling.*

This statement is a corollary of conclusions 1 and 2 above, according to which each parent molecule passes on two subunits to progeny molecules and each progeny molecule receives just one parental subunit.   It follows that each single molecular reproductive act results in a doubling of the number of molecules entering into that act.

The above conclusions are represented schematically in Figure 5.

The Watson-Crick Model.—A molecular structure for DNA has been proposed by Watson and Crick.[15] It has undergone preliminary refinement[16] without alteration of its main features and is supported by physical and chemical studies.[17] The structure consists of two polynucleotide chains wound helically about a common axis.   The nitrogen base (adenine, guanine, thymine, or cytosine) at each level

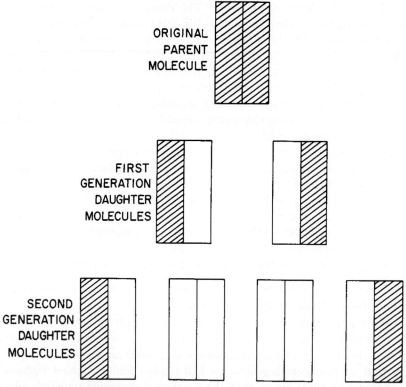

ORIGINAL PARENT MOLECULE

FIRST GENERATION DAUGHTER MOLECULES

SECOND GENERATION DAUGHTER MOLECULES

Fig. 5.—Schematic representation of the conclusions drawn in the text from the data presented in Fig. 4.   The nitrogen of each DNA molecule is divided equally between two subunits.   Following duplication, each daughter molecule receives one of these.   The subunits are conserved through successive duplications.

on one chain is hydrogen-bonded to the base at the same level on the other chain. Structural requirements allow the occurrence of only the hydrogen-bonded base pairs adenine-thymine and guanine-cytosine, resulting in a detailed complementariness between the two chains.   This suggested to Watson and Crick[18] a definite and structurally plausible hypothesis for the duplication of the DNA molecule. According to this idea, the two chains separate, exposing the hydrogen-bonding sites of the bases.   Then, in accord with the base-pairing restrictions, each chain serves as a template for the synthesis of its complement.   Accordingly, each daughter molecule contains one of the parental chains paired with a newly synthesized chain (Fig. 6).

678   *BIOLOGY: MESELSON AND STAHL*   Proc. N. A. S.

The results of the present experiment are in exact accord with the expectations of the Watson-Crick model for DNA duplication. However, it must be emphasized that it has not been shown that the molecular subunits found in the present experiment are single polynucleotide chains or even that the DNA molecules studied here correspond to single DNA molecules possessing the structure proposed by Watson and Crick. However, some information has been obtained about the molecules and their subunits; it is summarized below.

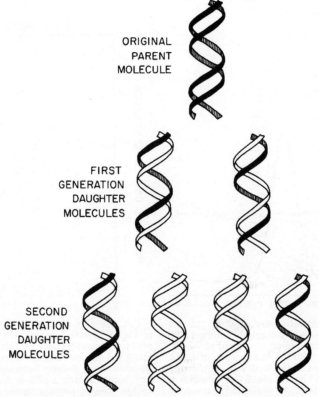

ORIGINAL
PARENT
MOLECULE

FIRST
GENERATION
DAUGHTER
MOLECULES

SECOND
GENERATION
DAUGHTER
MOLECULES

Fig. 6.—Illustration of the mechanism of DNA duplication proposed by Watson and Crick. Each daughter molecule contains one of the parental chains (*black*) paired with one new chain (*white*). Upon continued duplication, the two original parent chains remain intact, so that there will always be found two molecules each with one parental chain.

The DNA molecules derived from *E. coli* by detergent-induced lysis have a buoyant density in CsCl of 1.71 gm. cm.$^{-3}$, in the region of densities found for T2 and T4 bacteriophage DNA, and for purified calf-thymus and salmon-sperm DNA. A highly viscous and elastic solution of N^{14} DNA was prepared from a dodecyl sulfate lysate of *E. coli* by the method of Simmons[19] followed by deproteinization with chloroform. Further purification was accomplished by two cycles of preparative density-gradient centrifugation in CsCl solution. This purified bacterial DNA was found to have the same buoyant density and apparent molecular weight, 7×10^6, as the DNA of the whole bacterial lysates (Figs. 7, 8).

Heat Denaturation.—It has been found that DNA from *E. coli* differs importantly from purified salmon-sperm DNA in its behavior upon heat denaturation.

Exposure to elevated temperatures is known to bring about an abrupt collapse of the relatively rigid and extended native DNA molecule and to make available for acid-base titration a large fraction of the functional groups presumed to be blocked by hydrogen-bond formation in the native structure.[19, 20, 21, 22] Rice and Doty[22] have reported that this collapse is not accompanied by a reduction in molecular weight as determined from light-scattering. These findings are corroborated by density-gradient centrifugation of salmon-sperm DNA.[23] When this material is

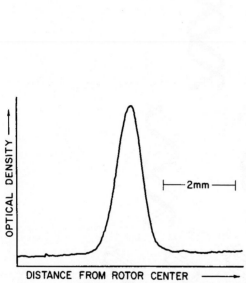

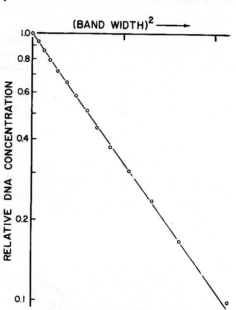

Fig. 7.—Microdensitometer tracing of an ultraviolet absorption photograph showing the optical density in the region of a band of N^{14} *E. coli* DNA at equilibrium. About 2 µg. of DNA purified as described in the text was centrifuged at 31,410 rpm at 25° in 7.75 molal CsCl at pH 8.4. The density gradient is essentially constant over the region of the band and is 0.057 gm./cm.[4]. The position of the maximum indicates a buoyant density of 1.71 gm. cm.[-3] In this tracing the optical density above the base line is directly proportional to the concentration of DNA in the rotating centrifuge cell. The concentration of DNA at the maximum is about 50 µg./ml.

Fig. 8.—The square of the width of the band of Fig. 7 plotted against the logarithm of the relative concentration of DNA. The divisions along the abscissa set off intervals of 1 mm.[2]. In the absence of density heterogeneity, the slope at any point of such a plot is directly proportional to the weight average molecular weight of the DNA located at the corresponding position in the band. Linearity of this plot indicates monodispersity of the banded DNA. The value of the the slope corresponds to an apparent molecular weight for the Cs·DNA salt of 9.4 × 10⁶, corresponding to a molecular weight of 7.1 × 10⁶ for the sodium salt.

kept at 100° for 30 minutes either under the conditions employed by Rice and Doty or in the CsCl centrifuging medium, there results a density increase of 0.014 gm. cm.[-3] with no change in apparent molecular weight. The same results are obtained if the salmon-sperm DNA is pre-treated at pH 6 with EDTA and sodium dodecyl sulfate. Along with the density increase, heating brings about a sharp reduction in the time required for band formation in the CsCl gradient. In the absence of an increase in molecular weight, the decrease in banding time must be ascribed[10] to an increase in the diffusion coefficient, indicating an extensive collapse of the native structure.

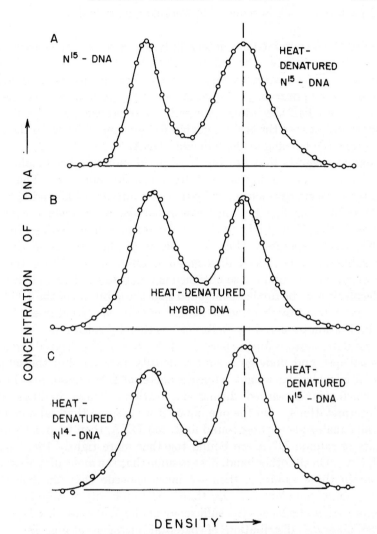

Fig. 9.—The dissociation of the subunits of *E. coli* DNA upon heat denaturation. Each smooth curve connects points obtained by microdensitometry of an ultraviolet absorption photograph taken after 20 hours of centrifugation in CsCl solution at 44,770 rpm. The baseline density has been removed by subtraction. *A:* A mixture of heated and unheated N^{15} bacterial lysates. Heated lysate alone gives one band in the position indicated. Unheated lysate was added to this experiment for comparison. Heating has brought about a density increase of 0.016 gm. cm. $^{-3}$ and a reduction of about half in the apparent molecular weight of the DNA. *B:* Heated lysate of N^{15} bacteria grown for one generation in N^{14} growth medium. Before heat denaturation, the hybrid DNA contained in this lysate forms only one band, as may be seen in Fig. 4. *C:* A mixture of heated N^{14} and heated N^{15} bacterial lysates. The density difference is 0.015 gm. cm. $^{-3}$

The decrease in banding time and a density increase close to that found upon heating salmon-sperm DNA are observed (Fig. 9, *A*) when a bacterial lysate containing uniformly labeled N^{15} or N^{14} *E. coli* DNA is kept at 100° C. for 30 minutes in the CsCl centrifuging medium; but the apparent molecular weight of the heated bacterial DNA is reduced to approximately half that of the unheated material.

Half-labeled DNA contained in a detergent lysate of N^{15} *E. coli* cells grown for one generation in N^{14} medium was heated at 100° C. for 30 minutes in the CsCl centrifuging medium. This treatment results in the loss of the original half-labeled material and in the appearance in equal amounts of two new density species, each with approximately half the initial apparent molecular weight (Fig. 9, *B*). The density difference between the two species is 0.015 gm. cm. $^{-3}$, close to the increment produced by the N^{15} labeling of the unheated DNA.

This behavior suggests that heating the hybrid molecule brings about the dissociation of the N^{15}-containing subunit from the N^{14} subunit. This possibility was tested by a density-gradient examination of a mixture of heated N^{15} DNA and heated N^{14} DNA (Fig. 9, *C*). The close resemblance between the products of heating hybrid DNA (Fig. 9 *B*) and the mixture of products obtained from heating N^{14} and N^{15} DNA separately (Fig. 9, *C*) leads to the conclusion that the two molecular subunits have indeed dissociated upon heating. Since the apparent molecular weight of the subunits so obtained is found to be close to half that of the intact molecule, it may be further concluded that the subunits of the DNA molecule which are conserved at duplication are single, continuous structures. The scheme for DNA duplication proposed by Delbrück[24] is thereby ruled out.

To recapitulate, both salmon-sperm and *E. coli* DNA heated under similar conditions collapse and undergo a similar density increase, but the salmon DNA retains its initial molecular weight, while the bacterial DNA dissociates into the two subunits which are conserved during duplication. These findings allow two different interpretations. On the one hand, if we assume that salmon DNA contains subunits analogous to those found in *E. coli* DNA, then we must suppose that the subunits of salmon DNA are bound together more tightly than those of the bacterial DNA. On the other hand, if we assume that the molecules of salmon DNA do not contain these subunits, then we must concede that the bacterial DNA molecule is a more complex structure than is the molecule of salmon DNA. The latter interpretation challenges the sufficiency of the Watson-Crick DNA model to explain the observed distribution of parental nitrogen atoms among progeny molecules.

Conclusion.—The structure for DNA proposed by Watson and Crick brought forth a number of proposals as to how such a molecule might replicate. These proposals[6] make specific predictions concerning the distribution of parental atoms among progeny molecules. The results presented here give a detailed answer to the question of this distribution and simultaneously direct our attention to other problems whose solution must be the next step in progress toward a complete understanding of the molecular basis of DNA duplication. What are the molecular structures of the subunits of *E. coli* DNA which are passed on intact to each daughter molecule? What is the relationship of these subunits to each other in a DNA molecule? What is the mechanism of the synthesis and dissociation of the subunits in vivo?

Summary.—By means of density-gradient centrifugation, we have observed the distribution of N^{15} among molecules of bacterial DNA following the transfer of a uniformly N^{15}-substituted exponentially growing *E. coli* population to N^{14} medium.

We find that the nitrogen of a DNA molecule is divided equally between two physically continuous subunits; that, following duplication, each daughter molecule receives one of these; and that the subunits are conserved through many duplications.

* Aided by grants from the National Foundation for Infantile Paralysis and the National Institutes of Health.

† Contribution No. 2344.

[1] R. D. Hotchkiss, in *The Nucleic Acids*, ed. E. Chargaff and J. N. Davidson (New York: Academic Press, 1955), p. 435; and in *Enzymes: Units of Biological Structure and Function*, ed. O. H. Gaebler (New York: Academic Press, 1956), p. 119.

[2] S. H. Goodgal and R. M. Herriott, in *The Chemical Basis of Heredity*, ed. W. D. McElroy and B. Glass (Baltimore: Johns Hopkins Press, 1957), p. 336.

[3] S. Zamenhof, in *The Chemical Basis of Heredity*, ed. W. D. McElroy and B. Glass (Baltimore: Johns Hopkins Press, 1957), p. 351.

[4] A. D. Hershey and M. Chase, *J. Gen. Physiol.*, **36**, 39, 1952.

[5] A. D. Hershey, *Virology*, **1**, 108, 1955; **4**, 237, 1957.

[6] M. Delbrück and G. S. Stent, in *The Chemical Basis of Heredity*, ed. W. D. McElroy and B. Glass (Baltimore: Johns Hopkins Press, 1957), p. 699.

[7] C. Levinthal, these Proceedings, **42**, 394, 1956.

[8] J. H. Taylor, P. S. Woods, and W. L. Huges, these Proceedings, **43**, 122, 1957.

[9] R. B. Painter, F. Forro, Jr., and W. L. Hughes, *Nature*, **181**, 328, 1958.

[10] M. S. Meselson, F. W. Stahl, and J. Vinograd, these Proceedings, **43**, 581, 1957.

[11] The buoyant density of a molecule is the density of the solution at the position in the centrifuge cell where the sum of the forces acting on the molecule is zero.

[12] Our attention has been called by Professor H. K. Schachman to a source of error in apparent molecular weights determined by density-gradient centrifugation which was not discussed by Meselson, Stahl, and Vinograd. In evaluating the dependence of the free energy of the DNA component upon the concentration of CsCl, the effect of solvation was neglected. It can be shown that solvation may introduce an error into the apparent molecular weight if either CsCl or water is bound preferentially. A method for estimating the error due to such selective solvation will be presented elsewhere.

[13] In addition to NH_4Cl, this medium consists of 0.049 M Na_2HPO_4, 0.022 M KH_2PO_4, 0.05 M NaCl, 0.01 M glucose, 10^{-3} M $MgSO_4$, and 3×10^{-6} M $FeCl_3$.

[14] This result also shows that the generation time is very nearly the same for all DNA molecules in the population. This raises the questions of whether in any one nucleus all DNA molecules are controlled by the same clock and, if so, whether this clock regulates nuclear and cellular division as well.

[15] F. H. C. Crick and J. D. Watson, *Proc. Roy. Soc. London*, A, **223**, 80, 1954.

[16] R. Langridge, W. E. Seeds, H. R. Wilson, C. W. Hooper, M. H. F. Wilkins, and L. D. Hamilton, *J. Biophys. and Biochem. Cytol.*, **3**, 767, 1957.

[17] For reviews see D. O. Jordan, in *The Nucleic Acids*, ed. E. Chargaff and J. D. Davidson (New York: Academic Press, 1955), **1**, 447; and F. H. C. Crick, in *The Chemical Basis of Heredity*, ed. W. D. McElroy and B. Glass (Baltimore: Johns Hopkins Press, 1957), p. 532.

[18] J. D. Watson and F. H. C. Crick, *Nature*, **171**, 964, 1953.

[19] C. E. Hall and M. Litt, *J. Biophys. and Biochem. Cytol.*, **4**, 1, 1958.

[20] R. Thomas, *Biochim. et Biophys. Acta*, **14**, 231, 1954.

[21] P. D. Lawley, *Biochim. et Biophys. Acta*, **21**, 481, 1956.

[22] S. A. Rice and P. Doty, *J. Am. Chem. Soc.*, **79**, 3937, 1957.

[23] Kindly supplied by Dr. Michael Litt. The preparation of this DNA is described by Hall and Litt (*J. Biophys. and Biochem. Cytol.*, **4**, 1, 1958).

[24] M. Delbrück, these Proceedings, **40**, 783, 1955.

Guiding Questions for Reading This Article

A. About the Article

1. State the last names of the authors of the article and the name of the university at which they worked.

2. What is the name of the scientific journal in which the article was published? In what year was it published?

3. This research was aided by grants from what organizations? (See the notes at the end of the article, page 682, before the references.)

4. Specialized vocabulary: Write a brief definition of each term.

   density-gradient centrifugation

   generation time

   isotope

B. About the Study

5. The investigators "anticipated that a label which imparts to the DNA molecule an increased density" (Introduction, paragraph 1) might help reveal how atoms in parental DNA are distributed to progeny DNA molecules. What label did the investigators use to be able to detect small density differences among molecules?

6. When bacteria cells are lysed (broken open) and the contents are placed in a cesium chloride mixture and spun in a centrifuge, the contents of the solution will be sorted by density. Figure 2a presents the results of separation of DNA composed of N^{14} from DNA composed of N^{15} by density-gradient centrifugation. The left side of the photograph is the part of the centrifuge tube closest to the axis of rotation (top of the tube), and the right side is where material has spun farthest from the center of rotation by centrifugal force (bottom of the tube). Which would have greater density, DNA composed of N^{14} or DNA

composed of N^{15}? Therefore, in a photograph such as in Figure 2a, which is the location of the higher-density material in each sample—to the left or to the right?

7. DNA absorbs ultraviolet light, so UV light photography can be used to make DNA visible. Figure 1 presents 13 images, each a photograph of ultraviolet absorption, indicating the presence of DNA at certain locations in the sample in a centrifuge tube. At time 0, before centrifugation, the dark band covers nearly the entire length of the sample in the tube, from left to right. This means that before centrifugation, the DNA molecules are scattered throughout the mixture of cell contents in the centrifuge tube and have not yet been sorted according to density. (a) After 10.7 hours of centrifugation, the dark band is limited to about what fraction of the length of the centrifuge tube? (b) After what amount of time (hours) in the centrifuge did the band begin to narrow? (c) After 36.5 hours of centrifugation, the dark band is located about where on the photograph—at the left, middle, or right?

8. Figure 2 shows two views of the same results. What is the relationship between the UV absorption bands shown in Figure 2a and the line with two peaks in Figure 2b?

9. In their experiment, how did the investigators cause the bacteria to switch from incorporating N^{15} into their DNA to growing with only light nitrogen isotopes, N^{14}?

10. Based on the second paragraph on p. 676, how does the relative position of a band indicate how much of the DNA was labeled with N^{15}?

11. (a) From the legend of Figure 3, what was the average bacterial generation time (in hours) for these experiments? (b) About how long would it take for four generations? (c) In Figure 3, what was the experimental change made at time 0?

12. Examine the results of Figure 4, in which higher density DNA is to the right. (a) What is the density difference between the initial DNA (0 generations) and the DNA after 1.0 generations? (b) How did the DNA density distribution after 1 generation compare to that after 2.5 generations? (c) DNA at high density (the farthest right band) is present at generation times 0, 0.3, and 0.7. At what *other* generations is the high-density DNA present?

13. The second point of the Discussion section (p. 676) states that each daughter molecule receives one subunit from the parent DNA. What pattern in the results supports this statement?

14. In the Discussion section, the authors draw connections between their own density results and the Watson-Crick proposal for a possible DNA duplication mechanism. In order for you to see the connections clearly, draw a quick sketch of three figures shown in this paper that represent the DNA after 1 generation: (a) the sketch from Figure 4, a tracing of the peak after 1.0 generation; (b) the schematic representation of the first-generation daughter molecules from Figure 5; and (c) the first-generation daughter molecules from Figure 6 (illustration of mechanism proposed by Watson and Crick).

15. Examine your sketches for (b) and (c) in the previous question. In each sketch, the two images are the same. If they were different from each other, how would the image in sketch (a) be different?

16. Summarize the "structurally plausible hypothesis" of Watson and Crick for the duplication of the DNA molecule, as explained on p. 677.

C. General Conclusions and Extensions of the Work

17. At the end of the paper, the authors state that the nitrogen of a DNA molecule is divided equally between two subunits and each daughter molecule receives one of these. (a) Which component of a DNA nucleotide contains nitrogen? (b) How could you label another part of the DNA and repeat these tests with another component?

18. In the introduction, the authors point out that hypotheses about DNA replication make different predictions about the distribution of parental atoms into daughter molecules. Today, we often refer to the model presented here as a "semi-conservative" model for replication, because components of each strand stay together (that is, the strand is conserved) but the two strands of the parent molecule do not stay together. What would happen in the distribution of parental atoms into daughter molecules if DNA replication followed a "conservative" model in which the entire parent DNA molecule was conserved?

19. The Meselson-Stahl experiments and this paper are considered important, groundbreaking studies in molecular biology. Decades later, the experiments are still admired, and "the paper is still held aloft for its clarity" [T. H. Davis, Meselson and Stahl: The art of DNA replication, *PNAS* 101:17895–17896 (2004).]. Some of the clarity of the centrifugation results was due to the fact that the DNA was fragmented during handling (although the experimenters did not know it at the time). As quoted in Davis (2004), Stahl said that pipetting DNA was like "throwing spaghetti over Niagara Falls." When DNA breaks into fragments, the sugar-phosphate backbone is broken, forming many short double helixes. Why did the fragmented DNA still show the same pattern of nitrogen density banding?

20. Imagine that you were a member of this research team and involved in these experiments. What could be a possible follow-up test that extends this work?

Inquiry Figure 23.16: *Do Females Select Mates Based on Traits Indicative of "Good Genes"?*

Introduction—The Article and Phenomenon Under Study

In many animals, males display to attract mates, and in some cases females are selective about the male with whom they mate. How do such female preferences for certain male characteristics evolve? How can we discover the genetic basis of evolution? By studying the gray tree frog, in which male vocalizations attract choosy females, investigators were able to examine the relationship between female choice and the genetic basis of evolution of male traits. This research, from the following paper, is the subject of Inquiry Figure 23.16 in *Campbell Biology*, Ninth Edition:

A. M. Welch et al., Call duration as an indicator of genetic quality in male gray tree frogs, *Science* 280:1928–1930 (1998).

Read the complete article beginning on the next page and then answer the questions following the article.

▼ **Figure 23.16**

INQUIRY

Do females select mates based on traits indicative of "good genes"?

EXPERIMENT Female gray tree frogs (*Hyla versicolor*) prefer to mate with males that give long mating calls. Allison Welch and colleagues, at the University of Missouri, tested whether the genetic makeup of long-calling (LC) males is superior to that of short-calling (SC) males. The researchers fertilized half the eggs of each female with sperm from an LC male and fertilized the remaining eggs with sperm from an SC male. The resulting half-sibling offspring were raised in a common environment, and several measures of their "performance" were tracked for two years.

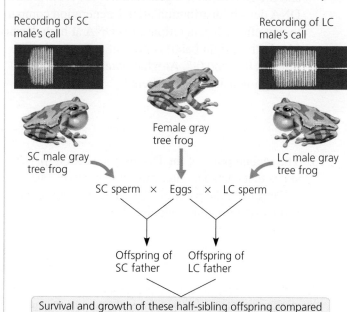

RESULTS

Offspring Performance	1995	1996
Larval survival	LC better	NSD
Larval growth	NSD	LC better
Time to metamorphosis	LC better (shorter)	LC better (shorter)

NSD = no significant difference; LC better = offspring of LC males superior to offspring of SC males.

CONCLUSION Because offspring fathered by an LC male outperformed their half-siblings fathered by an SC male, the team concluded that the duration of a male's mating call is indicative of the male's overall genetic quality. This result supports the hypothesis that female mate choice can be based on a trait that indicates whether the male has "good genes."

SOURCE A. M. Welch et al., Call duration as an indicator of genetic quality in male gray tree frogs, *Science* 280:1928–1930 (1998).

INQUIRY IN ACTION Read and analyze the original paper in *Inquiry in Action: Interpreting Scientific Papers*.

WHAT IF? Why did the researchers split each female frog's eggs into two batches for fertilization by different males? Why didn't they mate each female with a single male frog?

Call Duration as an Indicator of Genetic Quality in Male Gray Tree Frogs

Allison M. Welch,* Raymond D. Semlitsch, H. Carl Gerhardt

The "good genes" hypothesis predicts that mating preferences enable females to select mates of superior genetic quality. The genetic consequences of the preference shown by female gray tree frogs for long-duration calls were evaluated by comparing the performance of maternal half-siblings sired by males with different call durations. Offspring of male gray tree frogs that produced long calls showed better performance during larval and juvenile stages than did offspring of males that produced short calls. These data suggest that call duration can function as a reliable indicator of heritable genetic quality.

The "good genes" model of sexual selection predicts that some attributes of male courtship displays advertise genetic quality. Preferences for such attributes should allow females to mate with high-quality males and thereby benefit indirectly through enhanced quality of offspring (1). Although the good genes hypothesis has been tested several times (2), few studies have provided direct genetic evidence supporting this hypothesis (3). Only one such study involved a species in which females cannot benefit directly from their choice of mates (4). Because selection for direct benefits such as courtship feeding or parental care should overwhelm any selection for indirect (genetic) benefits (5), the role of good genes selection in the evolution and maintenance of female preferences is best tested in species in which females do not benefit directly from mate choice.

Female gray tree frogs (*Hyla versicolor*) strongly prefer male advertisement calls of long duration in laboratory experiments (6, 7). In the field, females freely initiate matings with calling males and do not always choose the first male encountered (7). Because males do not defend oviposition sites, offer nuptial gifts, or contribute parental care (8, 9), and no difference has been found in fertilization success as a function of call duration (10), there are no apparent direct benefits of a female's mate choice. We therefore predicted that females selecting mates with long calls should benefit indirectly

Division of Biological Sciences, University of Missouri, Columbia, MO 65211, USA.

*To whom correspondence should be addressed. E-mail: awelch@biosci.mbp.missouri.edu

through increased fitness of offspring. This prediction can be tested by evaluating the relation between paternal call duration and the genetic quality of offspring.

Male gray tree frog advertisement calls consist of rapidly repeated pulses. In dense choruses and in response to playbacks, males tend to increase call duration by increasing the number of pulses per call (11, 12). Nonetheless, some males consistently produce longer calls than others in the same acoustic environment (7, 12–14). Although long calls are usually produced at slow rates, thereby keeping aerobic metabolic costs relatively constant (11, 14), males that produce long calls spend less time calling per night (11) and attend fewer choruses per season (8) than males that produce short calls. Long calls thus appear to impose higher nonaerobic costs than short calls. Call duration may, therefore, be an honest indicator of male genetic quality.

We tested whether call duration indicates heritable genetic quality by using maternal half-siblingships (half-sibships) to compare the performance of different males' offspring while experimentally controlling for all maternal effects. Maternal half-sibships were generated by artificially crossing each female with two males that had been giving calls of distinctly different durations in the same social environment (Table 1). Thus, within each maternal half-sibship, one sibship was sired by a male with calls of longer duration than the male siring the other sibship. Because call duration varies with chorus density, males' calls must be assessed in the same social context in order to be validly compared. Thus, in 1995 we selected nine sets of two males that had been calling within 2 m of each other, and in 1996 we selected six sets of two field-caught males that had been calling simultaneously in a small captive chorus. The mean difference in call duration between the long- and the short-caller in each set was 10.1 pulses per call in 1995 and 15.8 pulses per call in 1996 (Table 1); in laboratory experiments, female *H. versicolor* routinely base preferences on differences of as few as 2 pulses per call (*15*). Furthermore, the average call durations of individual males classified as long-callers did not overlap with the average call durations of individuals classified as short-callers (*16*). Long- and short-callers did not differ in body mass. External artificial fertilization allowed unambiguous assignment of paternity, and rotation of the egg-stripping of each female between the pair of males eliminated the possibility of effects of fertilization order (*17*). In 1995, each of nine gravid females was artificially crossed with a different set of males to generate nine maternal half-sibships. In 1996, each of 11 gravid females was artificially crossed with at least one set of males to generate 16 maternal half-sibships. All frogs were collected near a pond in Boone County,

Missouri.

Because the relative performance of different genotypes can vary significantly with environmental conditions (*18*), we reared the resulting tadpoles at two food levels, thereby creating an unfavorable and a favorable growth environment in which to compare the performance of offspring (*19*). Comparison of our results with those from field studies indicates that our high food treatment was a realistic approximation of conditions encountered in nature (*20*). Tadpoles from the crosses (1995, $n = 538$; 1996, $n = 384$) were raised individually in containers filled with 1.0 liter of water in the laboratory at the two food levels; 15 tadpoles per family were reared at each food level in 1995 and six tadpoles per family in 1996 (*21*). To assess offspring performance, we used several variables (*22*) that are important determinants of fitness in anurans, predicting future survival and age and size at maturity, which influence lifetime reproductive success (*23*).

Offspring of males with long calls always performed significantly better than or not significantly differently from offspring of males with short calls (*24*) (Table 2). In multivariate analyses where responses were combined to account for correlations

among response variables (*25*), the main effect of call duration was significant at both food levels in 1996 and showed the same trend at both food levels in 1995 (Table 2), with offspring of males with long calls showing a general performance advantage over offspring of males with short calls. The probability of obtaining these four multivariate results that independently support the same directional hypothesis was calculated as $P = 0.0008$ (Table 2) with the use of a combined probability test (*26*). The multivariate tests therefore support the hypothesis that offspring performance is predicted by paternal call duration.

The specific benefits realized by offspring of long-callers differed among our experimental environments (Table 2). Because variation in the quality of the growth environment is predicted to influence the relation between larval growth and development (*27*), this difference in responses among environments is not unexpected. The consistency of the general benefit realized by offspring of long-callers in our experimental environments suggests that a general performance advantage may be applicable in other environments as well.

Overall, these results provide strong evidence that males with long calls relative to those of other males in the same social environment sired offspring of significantly higher phenotypic quality than males with short relative call durations. We can attribute these observed phenotypic differences to differences in paternal genetic contribution, because our comparison of maternal half-sibships controls for maternal genetic contributions and maternal effects. Thus, our results demonstrate that relative call duration reliably reflects genetic quality in *H. versicolor*. Our data suggest a genetic correlation between sire call duration and offspring performance, which implies that each trait has a heritable basis. The preference for long calls should, therefore, enable females to select high-quality mates and benefit indirectly through increased fitness of offspring. Because female *H. versicolor* do not gain direct benefits from their choice of mate, the indirect genetic benefits we have documented suggest good genes selection as a probable explanation for the evolution and maintenance of the female preference in this species.

Table 1. Average calling performance of sires exhibiting long versus short calls. For each 1995 male, approximately 25 consecutive calls were analyzed from field recordings. For each 1996 male, at least 20 min of consecutive calls were analyzed from digitally collected data.

Year	Performance	Call duration		Calling effort*
		Pulses per call	Seconds	
1995	Long-callers	28.3	1.74	0.214
	Short-callers	18.2	1.05	0.188
	Difference	10.1 ± 4.9†	0.69 ± 0.36†	0.026 ± 0.040‡
1996	Long-callers	31.5	1.41	0.092
	Short-callers	15.7	0.68	0.082
	Difference	15.8 ± 4.6†	0.72 ± 0.24†	0.010 ± 0.048‡

*Calling effort was measured as duty cycle—the proportion of time during which the individual was producing sound.   †$P < 0.001$; paired t test.   ‡$P > 0.05$; paired t test.

Table 2. Relative performance of offspring of males exhibiting long versus short calls. A shorter larval period is interpreted as better performance. For all other variables, larger values indicate better performance. NS, not significant. Dashes indicate data not collected in 1995.

Parameter	1995		1996	
	High food	Low food	High food	Low food
Larval growth	NS*	Long ≫ short†	Long ≫ short	Long > short‡
Larval period	Long ≫ short	NS	Long ≫ short	NS
Metamorphic mass	NS	Long > short	NS	NS
Larval survival	Long > short	NS	NS	NS
Postmetamorphic growth	–	–	NS	Long ≫ short
MANOVA	λ = 0.96, df = 3, $P = 0.0887$	λ = 0.81, df = 3, $P = 0.0590$	λ = 0.90, df = 4, $P = 0.0143$	λ = 0.71, df = 4, $P = 0.0216$
Combined probability test	$\chi^2 = 26.67$, df = 8, $P = 0.0008$			

*NS = $P > 0.05$; univariate ANOVA (*24*).   †Long ≫ short = $P < 0.01$.   ‡Long > short = $0.05 > P > 0.01$.

REFERENCES AND NOTES

1. A. Zahavi, *J. Theor. Biol.* **53**, 205 (1975).
2. R. D. Howard, H. H. Whiteman, T. I. Schueller, *Evolution* **48**, 1286 (1994); R. D. Semlitsch, *Behav. Ecol. Sociobiol.* **34**, 19 (1994); B. D. Woodward, *Am. Nat.* **128**, 58 (1986); S. L. Mitchell, *Evolution* **44**, 502 (1990).
3. K. Norris, *Nature* **362**, 537 (1993); B. C. Sheldon, J. Merilä, A. Qvarnström, L. Gustafsson, H. Ellegren,

Proc. R. Soc. London Ser. B **264**, 297 (1997).

4. M. Petrie, *Nature* **371**, 598 (1994).

5. M. Kirkpatrick and N. H. Barton *Proc. Natl. Acad. Sci. U.S.A.* **94**, 1282 (1997).

6. G. M. Klump and H. C. Gerhardt, *Nature* **326**, 286 (1987).

7. H. C. Gerhardt, M. L. Dyson, S. D. Tanner, *Behav. Ecol.* **7**, 7 (1996).

8. B. K. Sullivan and S. H. Hinshaw, *Anim. Behav.* **44**, 733 (1992).

9. G. M. Fellers, *Copeia* **1979**, 286 (1979).

10. J. D. Krenz, unpublished data.

11. K. D. Wells and T. L. Taigen, *Behav. Ecol. Sociobiol.* **19**, 9 (1986).

12. H. C. Gerhardt, *Anim. Behav.* **42**, 615 (1991).

13. L. S. Runkle, K. D. Wells, C. C. Robb, S. L. Lance, *Behav. Ecol.* **5**, 318 (1994).

14. U. Grafe, *Copeia* **1997**, 356 (1997).

15. H. C. Gerhardt and G. F. Watson, *Anim. Behav.* **50**, 1187 (1995); H. C. Gerhardt and S. D. Tanner, unpublished data.

16. In 1995, average numbers of pulses per call ranged from 15.3 to 21.8 for individual short-callers and from 22.1 to 38.7 for long-callers. In 1996, average pulse numbers ranged from 14.4 to 18.0 for short-callers and from 22.1 to 37.7 for long-callers.

17. R. D. Semlitsch, S. Schmiedehausen, H. Hotz, P. Beerli, *Evol. Ecol.* **10**, 531 (1996); R. D. Semlitsch, H. Hotz, G.-D. Guex, *Evolution* **51**, 1249 (1997).

18. D. S. Falconer and T. F. C. Mackay, *Introduction to Quantitative Genetics* (Longman, Essex, ed. 4, 1996).

19. Tadpoles were fed finely ground Tetra-Min fish flakes. The high food ration was typically as much as the tadpoles could consume and was always three times the low food ration. Larval survival, growth rate, length of larval period, and mass at metamorphosis differed significantly between the two food levels in both years ($P < 0.001$, t tests), demonstrating that the low food level constituted a significantly less favorable environment than the high food level.

20. Individuals metamorphosing from our high food treatment weighed an average of 0.359 g in 1995 and 0.319 g in 1996. Average mass at metamorphosis of *H. versicolor* reared in field enclosures was 0.340 g in the sun and 0.310 g in the shade (*13*). In a natural population of *H. chrysoscelis* (the cryptic sister species of *H. versicolor*), average mass at metamorphosis was 0.33 g (*28*).

21. Tadpoles were randomly assigned to food levels and blocks within our randomized block design. The experiment was conducted blind with respect to genetic identity.

22. Performance was measured as follows: wet mass (to the nearest milligram) on day 30 of the experiment—a measure of early larval growth rate; the larval period in days from the beginning of the experiment (stage 25) (*29*) to forelimb emergence (stage 42); and wet mass (to the nearest milligram) of metamorphs after tail resorption (stage 46). Larval survival was calculated as the proportion of individuals from each family surviving to metamorphosis. In 1996, postmetamorphic growth was calculated as the difference between wet mass achieved 30 days after metamorphosis and wet mass at metamorphosis.

23. K. A. Berven and D. E. Gill, *Am. Zool.* **23**, 85 (1983); D. C. Smith, *Ecology* **68**, 344 (1987).

24. Mixed-model univariate analyses of variance (ANOVAs) were performed for each response variable, at each food level during each year, to test for the main effects of call duration, maternal identity, paternal identity (1996 only), and blocking factors, as well as for interactions. In order to account for correlations between larval period and metamorphic mass, each was used as a covariate in the univariate analyses of the other. Metamorphic mass was also used as a covariate in analyses of postmetamorphic growth. ANOVA tables and more complete descriptions of analyses can be found at www.sciencemag.org/feature/data/976682.shl

25. A multivariate analysis of variance (MANOVA) was used for each food level during each year to test for multivariate effects of call duration, maternal identity, paternal identity (1996 only), and blocking factors. MANOVAs simultaneously included larval growth, larval period, metamorphic mass, and (in 1996) postmetamorphic growth but did not include survival because the unit of analysis for this variable was the family rather than the individual. We present results for the multivariate effect of call duration, based on Wilks' λ. Data were appropriately transformed in both univariate and multivariate analyses.

26. R. R. Sokal and F. J. Rohlf, *Biometry* (Freeman, San Francisco, 1981).

27. H. M. Wilbur and J. P. Collins, *Science* **182**, 1305 (1973); J. Travis, *Evolution* **44**, 502 (1984).

28. M. E. Ritke, J. G. Babb, M. K. Ritke, *J. Herpetol.* **24**, 135 (1990).

29. K. L. Gosner, *Herpetologica* **16**, 183 (1960).

30. We thank B. Buchanan and J. Schwartz for help in assessing males in 1996; J. Krenz for help with artificial crosses in 1996; A. Bullerdieck for assistance in raising tadpoles in 1996; M. Cherry, M. Cunningham, J. Krenz, M. Parris, A. Pomiankowski, T. Ryan, and J. Schwartz for comments on the manuscript; and the many people who helped collect frogs. This work was supported by an NSF predoctoral fellowship (A.M.W.), NSF and NIMH grants (H.C.G.), and a Sigma Xi Grant-in-aid of Research (A.M.W.).

12 December 1997; accepted 22 April 1998

Guiding Questions for Reading This Article

A. About the Article

1. What is the last name of the first author of this paper? The authors were affiliated with what university? Investigators on this team received financial support from what organizations? (See note 30, on p. 1930.)

2. Give the name of the journal and the year in which this paper was published.

3. Specialized vocabulary: Write a brief definition of each term.

   fitness (evolutionary)

   gravid

   half-sibship

   heritable

   multivariate analysis

   phenotype

   sexual selection

4. Give the common name and the scientific name (genus and species) of the frog studied in this article. How is the scientific name abbreviated in the body of the text?

B. About the Study

5. What does the "good genes" model of sexual selection predict about the role of male courtship displays?

6. In testing the "good genes" model, why was it significant to use a species in which females themselves do not benefit *directly* from choice of good mates?

7. In field observations of this species, what suggests that females exhibit mate choice? What observations of the males indicate that females receive no direct benefit from male choice during mating?

8. Some male frogs consistently produce calls of longer duration. What was noted about the behavior of males that produce long calls?

9. Why did the investigators think that long calls are potentially costly, and why did they decide that call duration might be an honest indicator of male genetic quality?

10. Why did the investigators use maternal half-sibships in testing whether male call duration indicated heritable genetic quality?

11. How many male frogs and how many female frogs were used in these experiments?

12. From Table 1 of the article, what was the difference in call duration, both in pulses per call and length in seconds, between long-callers and short-callers in 1995? According to the *statistical test information* in Table 1 (paired *t* test, p. 0.001), was the difference in pulses per call a significant difference? Based on *previous laboratory information* on female preferences (see top of p. 1929, at reference 15), was the difference in pulses per call in the 1995 study a meaningful difference?

13. What is "calling effort"? Based on Table 1, how did it differ in long-callers and short-callers?

14. How many tadpoles were raised and under what conditions? When rearing the offspring, why did investigators raise tadpoles in two food levels? How did investigators assign individual tadpoles to the various test groups? (See note 21 on p. 1930.)

15. What were some of the measures in offspring performance listed in Table 2? In cases in which offspring results differed between long-callers and short-callers, offspring of which of the two groups had better performance?

16. Is this an *observational study,* in which quantitative, observational data are taken but no experimental manipulation is made, or is this an *experimental study,* in which researchers make manipulations by which the effects of different variables are tested, one at a time?

17. Is this a *field study,* with data collected on organisms in their natural habitat, or is this a *lab study,* in which organisms are studied under controlled conditions in the laboratory?

C. General Conclusions and Extensions of the Work

18. What was the general conclusion about the relation of male call duration and offspring performance?

19. The authors conclude that males with relatively long calls sired offspring with higher-quality phenotypes. Based on that result, list the steps by which long male calls evolved by natural selection.

20. Gray tree frog females choose males based on their call duration. Given the offspring performance results from this study, explain how female choice and preference for long-caller males could have evolved by the process of natural selection.

21. Imagine that you were a member of this research team and involved in these experiments. What could be a possible follow-up test that extends this work? Briefly state another experiment or measurement you would do within this research system.

Inquiry Figure 37.14: *Does the Invasive Weed Garlic Mustard Disrupt Mutualistic Associations Between Native Tree Seedlings and Arbuscular Mycorrhizal Fungi?*

Introduction—The Article and Phenomenon Under Study

Many plants benefit from an underground mutualism with mycorrhizal fungi. One means by which non-native plants can disrupt biological communities is by interfering with such a mutualism. Inquiry Figure 37.14 in *Campbell Biology, Ninth Edition,* presents experimental data testing the possibility of such interference from the following paper:

K. A. Stinson et al., Invasive plant suppresses the growth of native tree seedlings by disrupting belowground mutualisms, *PLoS Biol* (*Public Library of Science: Biology*) 4(5): e140 (2006).

Read the complete article beginning on the next page and then answer the questions following the article.

▼ Figure 37.14 **INQUIRY**

Does the invasive weed garlic mustard disrupt mutualistic associations between native tree seedlings and arbuscular mycorrhizal fungi?

EXPERIMENT Kristina Stinson, of Harvard University, and colleagues investigated the effect of invasive garlic mustard on the growth of native tree seedlings and associated mycorrhizal fungi. In one experiment, they grew seedlings of three North American trees—sugar maple, red maple, and white ash—in four different soils. Two of the soil samples were collected from a location where garlic mustard was growing, and one of these samples was sterilized. The other two soil samples were collected from a location devoid of garlic mustard, and one was then sterilized.  After four months of growth, the researchers harvested the shoots and roots and determined the dried biomass. The roots were also analyzed for percent colonization by arbuscular mycorrhizal fungi.

RESULTS Native tree seedlings grew more slowly and were less able to form mycorrhizal associations when grown either in sterilized soil or in unsterilized soil collected from a location that had been invaded by garlic mustard.

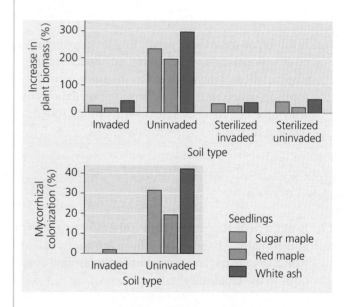

CONCLUSION The data support the hypothesis that garlic mustard suppresses growth of native trees by affecting the soil in a way that disrupts mutualistic associations between the trees and arbuscular mycorrhizal fungi.

SOURCE K. A. Stinson et al., Invasive plant suppresses the growth of native tree seedlings by disrupting belowground mutualisms, *PLoS Biol* (*Public Library of Science: Biology*) 4(5): e140 (2006).

INQUIRY IN ACTION Read and analyze the original paper in *Inquiry in Action: Interpreting Scientific Papers.*

WHAT IF? What effect would applying inorganic phosphate to soil invaded by garlic mustard have on the plant's ability to outcompete native species?

OPEN ACCESS Freely available online

PLoS BIOLOGY

Invasive Plant Suppresses the Growth of Native Tree Seedlings by Disrupting Belowground Mutualisms

Kristina A. Stinson[1], Stuart A. Campbell[2], Jeff R. Powell[2], Benjamin E. Wolfe[2], Ragan M. Callaway[3], Giles C. Thelen[3], Steven G. Hallett[4], Daniel Prati[5], John N. Klironomos[2*]

1 Harvard Forest, Harvard University, Petersham, Massachusetts, United States of America, 2 Department of Integrative Biology, University of Guelph, Guelph, Ontario, Canada, 3 Division of Biological Sciences, University of Montana, Missoula, Montana, United States of America, 4 Department of Botany and Plant Pathology, Purdue University, West Lafayette, Indiana, United States of America, 5 Department of Community Ecology, UFZ Centre for Environmental Research, Halle, Germany

The impact of exotic species on native organisms is widely acknowledged, but poorly understood. Very few studies have empirically investigated how invading plants may alter delicate ecological interactions among resident species in the invaded range. We present novel evidence that antifungal phytochemistry of the invasive plant, *Alliaria petiolata*, a European invader of North American forests, suppresses native plant growth by disrupting mutualistic associations between native canopy tree seedlings and belowground arbuscular mycorrhizal fungi. Our results elucidate an indirect mechanism by which invasive plants can impact native flora, and may help explain how this plant successfully invades relatively undisturbed forest habitat.

Citation: Stinson KA, Campbell SA, Powell JR, Wolfe BE, Callaway RM, et al. (2006) Invasive plant suppresses the growth of native tree seedlings by disrupting belowground mutualisms. PLoS Biol 4(5): e140. DOI: 10.1371/journal.pbio.0040140

Introduction

Widespread anthropogenic dispersal of exotic organisms has raised growing concern over their devastating ecological impacts, and has prompted decades of research on the ecology of invasive species [1–3]. Exotic plants may become aggressive invaders outside their home ranges for a number of reasons, including release from native, specialized antagonists [4], higher relative performance in a new site [5], direct chemical (allelopathic) interference with native plant performance [6], and variability in the responses and resistance of native systems to invasion [7,8]. Thus, successful invasion in many cases appears to involve the fact that invasive species are not at equilibrium, and are either freed of long-standing biotic interactions with their enemies in the home range, and/or disrupt interactions among the suite of native organisms they encounter in a new range [9]. Nevertheless, experimental data on species-level impacts of exotic plants are still limited [10]. One particularly understudied area is the potential for invasive plants to disrupt existing ecological associations within native communities [6,10]. Many exotic and native plants alike depend upon mutualisms with native insects, birds, or mammals for pollination and seed dispersal [11], and with soil microbes for symbiotic nutrient exchange [12]. Thus, when an introduced species encounters a new suite of resident organisms, it is likely to alter closely interlinked ecological relationships, many of which have co-evolved within native systems [6,11].

One such relationship is that between plants and mycorrhizal fungi [12]. Most vascular plants form mycorrhizal associations with arbuscular mycorrhizal fungi (AMF) [12], and many plants are highly dependent on this association for their growth and survival [12], particularly woody perennials and others found in late-successional communities [13]. In contrast, many weedy plants, in particular non-mycotrophic

plants, can be negatively affected by AMF [14–16]. Naturalized exotic plants have been found to be poorer hosts and depend less on native AMF than native plants [17]. They often colonize areas that have been disturbed [2], and disturbances to soil have been shown to negatively impact AMF functioning [18]. Furthermore, it has been proposed that the proliferation of plants with low mycorrhizal dependency may degrade AMF densities in the soil [17]. However, a few invasive plants proliferate in the understory of mature temperate forests [2], where AMF density is typically high [19]. The existing mycelial network in mature forest soils may facilitate the establishment of exotic, mycorrhizal-dependent, recruits [20,21], but this should not be the case for non-mycorrhizal invaders. If non-mycorrhizal invasive plants establish and degrade AMF in mature forests, then the effects on certain resident native plants could be substantial.

One of the most problematic invaders of mesic temperate forests in North America is *Alliaria petiolata* (garlic mustard; Brassicaceae), a non-mycorrhizal, shade-tolerant, Eurasian biennial herb which, like most other mustards, primarily occupies disturbed areas. Garlic mustard is abundant in forest edges, semishaded floodplains, and other disturbed sites in its home range [22]. However, this species has recently become an aggressive and widespread invader of both

Academic Editor: Michel Loreau, McGill University, Canada

Received December 5, 2005; Accepted March 1, 2006; Published April 25, 2006

DOI: 10.1371/journal.pbio.0040140

Abbreviations: AMF, arbuscular mycorrhizal fungi; ANOVA, analysis of variance; REGW, Ryan-Einot-Gabriel-Welsch

* To whom correspondence should be addressed. E-mail: jklirono@uoguelph.ca

disturbed areas and closed-canopy forest understory across much of the United States and Canada [23], where it apparently suppresses native understory plants, including the seedlings of dominant canopy trees [22,24]. The mechanism underlying garlic mustard's unusual capacity to enter and proliferate within intact North American forest community has not yet been established.

As shown in recent greenhouse experiments, garlic mustard's impact on native understory flora may involve competitive [25] or allelopathic effects on native plants [26], but it has also been hypothesized that this species interferes with plant–AMF interactions in its invaded range [27]. Members of the Brassicaceae, including garlic mustard, produce various combinations of glucosinolate products [28], organic plant chemicals with known anti-herbivore, anti-pathogenic and allelopathic [29] properties, that may also prevent this non-mycorrhizal plant family from associating with AMF [30]. These phytochemicals may be released into soils as root exudates, as a result of damaged root tissue, or in the form of leaf litter. High densities of garlic mustard in the field correlate with low inoculum potential of AMF, and extracts of garlic mustard leaves have been shown to reduce the germination of AMF spores and impair AMF colonization of cultivated tomato roots in laboratory settings [27]. Although not all Brassicaceae are invasive, it is possible that garlic mustard's successful invasion of understory habitats involves the negative effects of its phytochemistry on the native plant and AMF species it encounters outside its home range. Others have shown that exotic plants can recruit different suites of microbial organisms in their new ranges that can be antagonistic to native plants [6]. However, to our knowledge, no previous studies have directly tested whether this species or any other exotic plant disrupts native plant–AMF mutualisms within natural communities. Here, we present novel evidence that garlic mustard negatively impacts the growth of AMF-dependent forest tree seedlings by its disruption of native mycorrhizal mutualisms. We further show that, because seedlings of dominant tree species in mature forest communities are more highly dependent on AMF than plants that typically dominate earlier successional communities, garlic mustard invasion may disproportionately damage mature forests relative to other habitats.

Results/Discussion

We first tested whether native tree seedlings were less able to form mycorrhizal associations when grown in forest understory soils with a history of garlic mustard invasion than when grown in soils that had not experienced invasions (Experiment 1). We found that dominant native hardwood tree species of northeastern temperate forests, *Acer saccharum* (sugar maple), *Ac. rubrum* (red maple), and *Faximus americana* (white ash), showed significantly less AMF colonization of roots (Figure 1A) and slower growth (Figure 1B) when grown in soil that had been invaded by garlic mustard. AMF colonization was almost undetectable in soil that had been invaded by garlic mustard. These reductions were similar to those observed when seedlings were grown in sterilized soil from both garlic mustard–invaded and garlic mustard–free sites (Figure 1B), strongly suggesting that the mechanism by which garlic mustard suppresses the growth of native tree

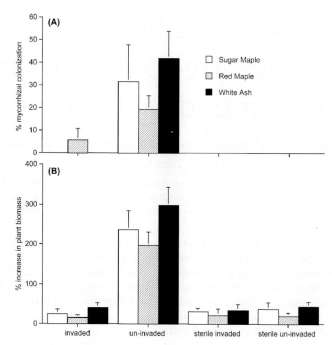

Figure 1. Experiment 1

The influence of field soils that were invaded or uninvaded by *Al. petiolata* (± sterilized) on (A) mycorrhizal colonization ($F_{sugar\ maple} = 77.7$, $df = 3,39$, $p < 0.001$; $F_{red\ maple} = 60.5$, $df = 3,39$, $p < 0.001$; and $F_{white\ ash} = 116.6$, $df = 3,39$, $p < 0.001$) and (B) biomass accumulation ($F_{sugar\ maple} = 57.8$, $df = 3,39$, $p < 0.001$; $F_{red\ maple} = 61.4$, $df = 3,39$, $p < 0.001$; and $F_{white\ ash} = 70.1$, $df = 3,39$, $p < 0.001$) of native tree seedlings. Bars represent the mean and standard error.
DOI: 10.1371/journal.pbio.0040140.g001

species is microbially-mediated, and not the result of soil differences or direct allelopathy.

We then conducted additional experiments to confirm that garlic mustard specifically caused AMF decline in the native soils (Experiment 2–4). We grew seedlings of the same three native tree species used in Experiment 1 in uninvaded forest soils that were conditioned for 3 mo with either garlic mustard plants or with one of the three native tree species. All three tree species demonstrated significantly lower AMF colonization in soils conditioned by *Al. petiolata* (0%–10%) than in soils conditioned by the native plants (20%–65%; Figure 2A). AMF colonization was similar in unconditioned (control) soils and soils conditioned with native plants. In addition, growth of the tree seedlings was the lowest in soils conditioned by garlic mustard (Figure 2B), confirming that garlic mustard plants reduce native plant performance by interfering with the formation of mycorrhizal associations.

We investigated whether there is a phytochemical basis to garlic mustard's observed antifungal effects on AMF in Experiments 3–4. In an earlier study, Vaughn and Berhow [31] isolated the phytotoxic glucosinolate hydrolysis products allyl isothiocyanate, benzyl isothiocyanate, and glucotropaeolin from extracts of *Al. petiolata* root tissues and found evidence for their allelopathic effects on certain plants in the absence of mycorrhizas. These phytochemicals could have direct effects on plant growth through allelopathy as well as indirect effects via disruption of AMF. To experimentally establish that garlic mustard's effect on AMF is phytochemically based, we grew native tree seedlings on uninvaded soils to which we added individual aqueous extracts of garlic

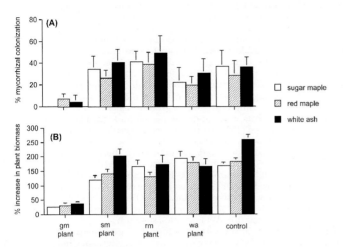

Figure 2. Experiment 2

The effect of soils conditioned with garlic mustard *Al. petiolata* (gm), sugar maple (sm), red maple (rm), or white ash (wa) on (A) mycorrhizal colonization ($F_{sugar\ maple} = 31.2$, $df = 4,49$, $p < 0.001$; $F_{red\ maple} = 18.2$, $df = 4,49$, $p < 0.001$; and $F_{white\ ash} = 22.1$, $df = 4,49$, $p < 0.001$) and (B) increase in biomass ($F_{sugar\ maple} = 15.1$, $df = 4,49$, $p < 0.001$; $F_{red\ maple} = 18.1$, $df = 4,49$, $p < 0.001$; and $F_{white\ ash} = 13.2$, $df = 4,49$, $p < 0.001$) of native tree seedlings. Bars represent the mean and standard error.
DOI: 10.1371/journal.pbio.0040140.g002

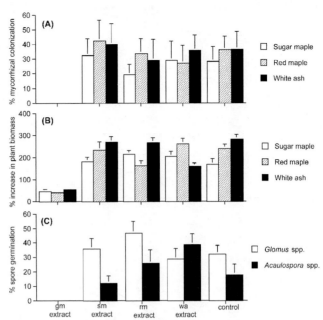

Figure 3. Experiments 3 and 4

The effects of extract of garlic mustard (gm), sugar maple (sm), red maple (rm), white ash (wa), or a water control on (A) mycorrhizal colonization of native tree seedlings ($F_{sugar\ maple} = 20.3$, $df = 4,49$, $p < 0.001$; $F_{red\ maple} = 19.8$, $df = 4,49$, $p < 0.001$; and $F_{white\ ash} = 25.4$, $df = 4,49$, $p < 0.001$ [Experiment 3]), (B) increase in biomass of native tree seedlings ($F_{sugar\ maple} = 11.7$, $df = 4,49$, $p < 0.001$; $F_{red\ maple} = 14.2$, $df = 4,49$, $p < 0.001$; and $F_{white\ ash} = 27.9$, $df = 4,49$, $p < 0.001$ [Experiment 3]), and (C) percent germination of native AMF spores ($F_{Glomus} = 17.3$, $df = 4,49$, $p < 0.001$; and $FA_{caulospora} = 21.8$, $df = 4,49$, $p < 0.001$ [Experiment 4]). Bars represent the mean and standard error.
DOI: 10.1371/journal.pbio.0040140.g003

mustard or each of the native trees species (Experiment 3). We found that garlic mustard extract was just as effective as the living plant at reducing AMF colonization (Figure 3A) and growth (Figure 3B) of the native plants. Moreover, exposing AMF spores to extract of garlic mustard severely and significantly reduced germination rates of those spores (Experiment 4; Figure 3C). Collectively, our results clearly demonstrate that garlic mustard, probably through phytochemical inhibition, disrupts the formation of mycorrhizal associations. Our results thus reveal a powerful, indirect mechanism by which an invasive species can suppress the growth of native flora.

Because plants vary in their dependency on AMF [32], garlic mustard's disruption of native plant–fungal mutualisms should not inhibit the growth of all plants equally, but rather should correlate strongly with the mycorrhizal dependence of species encountered in the invaded range. Specifically, courser root production, which impedes the nutrient uptake of typically slow-growing, woody plants such as tree seedlings, may explain the stronger AMF dependency of certain species [19,33]. To test whether garlic mustard's effects correlate with AMF dependency, and whether garlic mustard has stronger negative effects on forest tree seedlings than on other plants, we conducted another experiment (Experiment 5) using 16 plant species for which we determined AMF-dependency by computing the difference in plant growth in the presence and absence of AMF. We then tested the impact of garlic mustard on the AM fungal colonization and growth of each plant species as above. All 16 plants were successfully colonized by AMF, and the presence of garlic mustard heavily reduced AMF colonization in all plants (Figure 4A). However, the presence of garlic mustard had a much stronger effect on plants that had high mycorrhizal dependency than those with less dependency (Figure 4B). The strongest effects were observed for woody species most typically found in forested sites. These results indicate that the invasion of garlic mustard is more likely to negatively impact highly mycor-

rhizal-dependent tree seedlings than less-mycorrhizal-dependent plants. Thus, garlic mustard's successful colonization of understory habitat may be attributed in part to its ability to indirectly suppress woody competitors, and its effect on the native flora may be more detrimental in intact forests than disturbed sites. In addition, the data suggest that invasion by garlic mustard may have profound effects on the composition of mature forest communities (e.g., by repressing the regeneration of dominant canopy trees, and by favoring plants with low mycorrhizal dependency such as weedy herbs).

In conclusion, our results reveal a novel mechanism by which an invasive plant can disrupt native communities: by virtually eliminating the activity of native AMF from the soil and drastically impairing the growth of native canopy species. It is currently unclear precisely which phytochemicals produced by garlic mustard have the observed antifungal properties, whether and how they interact with other soil microbes, and whether these anti-fungal effects extend to other functionally important forest soil fungi such as ectomycorrhizal fungi and saprotrophic fungi. In addition, within the home range, it is not known if evolutionary natural resistance of co-occurring European neighbors may buffer the effects of garlic mustard's antifungal properties [34–36]. Further research in these directions is needed to better understand the effects of this invader on natural ecosystems and the mechanisms involved. In North America; however, the disruption of native tree seedling–AMF mutualisms may facilitate garlic mustard's invasion into mature forest under-

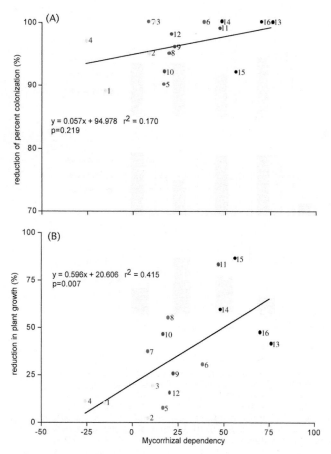

Figure 4. Experiment 5

(A) Effect of mycorrhizal dependency on *Al. petiolata* reduction of AMF colonization.

(B) Effect of mycorrhizal dependency on *Al. petiolata* reduction in plant growth. Mycorrhizal dependency was calculated separately as the difference between plant growth in the presence and absence of AMF. Different colors represent plants with different life-history strategies, as follows: yellow dot, herbaceous colonizers of disturbed edges and bare ground; reddish brown dot, herbaceous edge and gap species; blue dot, woody colonizers of forest edges and gaps; black dot, tree species of mature forest. Species are labeled as follows (with mean mycorrhizal colonization in soil not conditioned by garlic mustard ± standard error in parentheses): 1 = *Ci. intybus* (18.5 ± 4.1), 2 = *Tr. repens* (46.7 ± 6.3), 3 = *Pl. major* (28.2 ± 3.7), 4 = *Ta. officinale* (37.3 ± 2.5), 5 = *S. canadensis* (48.0 ± 6.2), 6 = *C. leucanthemum* (34.6 ± 3.1), 7 = *D. carota* (40.4 ± 6.2), 8 = *As. syriaca* (52.1 ± 5.8), 9 = *J. virginiana* (31.2 ± 4.4), 10 = *Po. deltoids* (63.9 ± 4.5), 11 = *M. alba* (38.6 ± 5.9), 12 = *Pr. virginiana* (28.4 ± 4.2), 13 = *Fr. americana* (65.9 ± 5.3), 14 = *Ac. saccharum* (46.3 ± 3.7), 15 = *Ac. rubrum* (59.5 ± 5.7), 16 = *Pr. serotina* (34.8 ± 5.5).

DOI: 10.1371/journal.pbio.0040140.g004

story and have particularly negative effects on the growth, survival, and recruitment of native trees, and the composition of forest communities.

Materials and Methods

Experiment 1. Using a 15-cm–wide corer, we collected soil from garlic mustard–invaded and nearby garlic mustard–free locations at each of five forested areas dominated by *Acer rubrum* L. (red maple), *Ac. saccharum* Marsh. (sugar maple), *Fraxinus americana* L. (white ash), and *Fagus grandifolia* Ehrh. (American beech) near Waterloo, Ontario, Canada. Invaded and uninvaded sites were randomly chosen within a 40-m² plot within each forested area. Soils from the invaded and uninvaded areas were pooled separately in the lab and screened to remove coarse roots and debris. Half the soil from each pool was then sterilized by autoclaving at 120 °C to create four soil treatments: (1) soil with a history of garlic mustard, (2) sterile soil with a history of

garlic mustard, (3) soil without a history of garlic mustard, and (4) sterile soil without a history of garlic mustard. Six-inch pots were filled with a 1:1 mixture of sterilized silica sand and one of the four soil types. To each pot, we added a single seedling (seeds germinated on Turface [Aimcor, Buffalo Grove, Illinois, United States], a clay substrate) of one of the three native overstory tree species (sugar maple, red maple, or white ash) in a complete 4 × 3 factorial design with ten replicates of each treatment combination. The initial wet biomass of each seedling was recorded prior to planting, and dry weights were estimated using a dry–wet regression calculated from twenty extra seedlings. Pots were randomly placed on a greenhouse bench. Plants were watered (400 ml) once per week. Fertilizer was not added. After 4 mo of growth, shoots and roots were harvested, dried at 60 °C for 48 h, and weighed to determine biomass. An approximately 1-g subsample of roots from each seedling was extracted, stained with Chlorazol Black E [37] and analyzed for percent colonization by AMF [38]. Biomass and percent colonization data were analyzed using analysis of variance (ANOVA) for two fixed effects (soil type and species) and their interaction, followed by the Ryan-Einot-Gabriel-Welsch (REGW) multiple-range test.

Experiment 2. Using field soil without a history of garlic mustard invasion (see Experiment 1), we grew garlic mustard, sugar maple, red maple, and white ash seedlings in separate 6-in pots ($n = 10$) to condition the soil to each plant species. After 3 mo of conditioning, shoots and roots were removed. Unconditioned soil served as a control to the four plant-conditioning treatments. We added a single seedling of each of the three tree species to each of the five soil treatments. Pots were randomly placed on a greenhouse bench. Plants were watered (400 ml) once per week, without fertilizer. After 4 mo of growth, plants were harvested, biomass was determined, and percent mycorrhizal colonization of roots was assessed as in Experiment 1. Data were analyzed using ANOVA for two fixed effects (species and soil condition treatment). Means from the three species were pooled, and the effect of conditioning treatment was tested with a single-factor ANOVA followed by the REGW multiple-range test.

Experiment 3. To 6-in pots containing field soil without a history of garlic mustard (see Experiment 1), we added a one-time, 100-ml aqueous extract [27] of whole plants of either garlic mustard, sugar maple, red maple, or white ash. A water control was included to give five treatments. Whole-plant extract was used to account for secondary compounds exuded through roots and leaf litter. After 1 wk of exposure to the extract, seedlings of each tree species were planted in each of these five treatments to give a full factorial design (extract source × tree species) with ten replicates of each treatment combination. Plants were watered (40 ml) every week, without fertilizer. After 4 mo of growth, plants were harvested, biomass was determined, and roots were assayed for mycorrhizal colonization as in Experiment 1. Data were analyzed by two-factor ANOVA.

Experiment 4. Spores from AMF native to the forest sites were obtained using trap cultures (as described in [39], but with a mix of native plants) of soil samples from the uninvaded locations. We visually collected and separated *Glomus* and *Acaulospora* spores from these cultures, and compared germination rates of each genus in five treatments: a water agar control and water agar amended with an aqueous extract from each of the four plants, as above. Ten randomly drawn spores were added into each plate, which was then incubated at 18 °C for 10 d. Ten replicate plates were prepared for each of the ten treatment combinations (two AMF genera × five extracts). Plates were monitored microscopically for spore germination. Percent germination data were analyzed using ANOVA for two fixed effects (extract source and AMF genus), and because of a significant interaction, each AMF genus was then analyzed separately using single-factor ANOVA followed by the REGW multiple-range test.

Experiment 5. We investigated the effects of garlic mustard on woody and herbaceous plants using the following 16 native plant species: *Cichorium intybus, Trifolium repens, Plantago major,* and *Taraxacum officinale* (dominant herbaceous colonizers of disturbed edges and bare ground); *Solidago canadensis, Chrysanthemum leucanthemum, Daucus carota,* and *Asclepias syriaca* (dominant herbaceous edge and gap species); *Juniperus virginiana, Populus deltoides, Morus alba,* and *Prunus virginiana* (dominant woody colonizers of forest edges and gaps); and *Fr. americana, Ac. saccharum, Ac. rubrum,* and *Pr. serotina* (dominant tree species of mature forest). Seedlings of each plant were transplanted into 8-in pots. For each species, growth was compared under the following soil treatments: (1) soil without a history of garlic mustard and inoculated with AMF, (2) soil without a history of garlic mustard, without AMF, and (3) soil with a history of garlic mustard, and inoculated with AMF. Experimental soil was collected within a mature-canopy maple forest from locations with and without garlic mustard. Soils from each location type were then mixed, cleaned of all coarse roots and debris, autoclaved, and added to the pots as a 1:1 mix of soil and silica sand. AMF spores were extracted from field soil collected from sites representing the four different habitats, and pooled. The AMF-inoculation treatment consisted of adding 200 randomly picked spores to each pot, 2 cm below the surface, and beneath the newly transplanted seedlings. Plants were watered (500 ml) once per week, without fertilizer. They were harvested after 4 mo of growth, dried at 60 °C for 36 h, and weighed to determine biomass. AMF dependency of each plant species was determined by computing the difference in plant growth in the presence and absence of AMF, i.e., contrast of treatments (1) and (2) [32]. The effects of garlic mustard on plant growth and percent colonization of each plant were determined by contrasting treatments (1) and (3). To ask whether any relationships existed among mycorrhizal dependency, life form, and garlic mustard effects, we performed two regressions: percent reduction in AMF colonization by garlic mustard on AMF dependency and percent reduction in plant biomass by garlic mustard on AMF dependency.

Acknowledgments

We thank T. Denich, V. Grebogi, G. Herrin, P. Hudson, G. Kuenen, J. Lozi, B. Shelton, P. Stephens, J. Van Houten, and Z. Zhu for technical assistance, and P. Antunes, G. De Deyn, and M. Hart for helpful comments on the text.

Author contributions. KAS, RMC, and JNK conceived and designed the experiments. KAS and JNK performed the experiments. KAS, SAC, JRP, BEW, RMC, GCT, SGH, DP, and JNK analyzed the data. JNK contributed reagents/materials/analysis tools. All authors wrote the paper.

Funding. We thank the Natural Sciences and Engineering Research Council of Canada, and the Harvard University Bullard Foundation for financial support.

Competing interests. The authors have declared that no competing interests exist.   ∎

References

1. Rejmánek M (2000) Invasive plants: Approaches and predictions. Austral Ecol 25: 497–506.
2. Mooney HA, Hobbs RJ (2000) Invasive species in a changing world. Washington (D. C.): Island Press. 457 p.
3. Ewel JJ, O'Dowd DJ, Bergelson J, Daehler CC, D'Antonio CM, et al. (1999) Deliberate introductions of species: Research needs. Bioscience 49: 619–630.
4. Mitchell CE, Power AG (2003) Release of invasive plants from viral and fungal pathogens. Nature 421: 625–627.
5. Thébaud C, Simberloff D (2001) Are plants really larger in their introduced ranges? Am Nat 157: 231–236.
6. Callaway RM, Ridenour WM (2004) Novel weapons: Invasive success and the evolution of increased competitive ability. Front Ecol Environ 2: 436–443.
7. Hobbs RJ, Huenneke LF (1992) Disturbance, diversity, and invasion: Implications for conservations. Conserv Biol 6: 324–337.
8. Levine JM, D'Antonio CM (1999) Elton revisited: A review of evidence linking diversity and invasibility. Oikos 87: 15–26.
9. Keane RM, Crawley MJ (2002) Exotic plant invasions and the enemy release hypothesis. Trends Ecol Evol 17: 164–170.
10. Levine JM, Vila M, D'Antonio CM, Dukes JS, Grigulis K, et al. (2003) Mechanisms underlying the impacts of exotic plant invasions. Proc Biol Sci 270: 775–781.
11. Richardson DM, Allsopp N, D'Antonio CM, Milton SJ, Rejmánek M (2000) Plant invasions—The role of mutualisms. Biol Rev Camb Philos Soc. 75: 65–93.
12. Smith SE, Read DJ (1997) Mycorrhizal symbiosis. 2nd edition. New York: Academic Press. 605 p.
13. Janos DP (1980) Mycorrhizae influence tropical succession. Biotropica 12: 56–64.
14. Grime JP, Mackey JML, Hillier SH, Read DJ (1987) Floristic diversity in a model system using experimental microcosms. Nature 328: 420–422.
15. Francis R, Read DJ (1995) Mutualism and antagonism in the mycorrhizal symbiosis, with special reference to impacts on plant community structure. Can J Bot 73: 1301–1309.
16. Van der Heijden MGA, Klironomos JN, Ursic M, Moutoglis P, Streitwolf-Engel R, et al. (1998) Mycorrhizal fungal diversity determines plant biodiversity, ecosystem variability and productivity. Nature 396: 69–72.
17. Vogelsang KM, Bever JD, Griswold M, Schultz PA (2004 June) The use of mycorrhizal fungi in erosion control applications. Final Report for Caltrans. Sacramento (California): California Department of Transportation Contract No. 65A0070. 150 p.
18. Haselwandter K (1997) Soil micro-organisms, mycorrhiza, and restoration ecology. In: Urbanska KM, Webb NR, Edwards PJ, editors. Restoration ecology and sustainable development. Cambridge: Cambridge University Press. pp. 65–80.
19. Read DJ (1991) Mycorrhizas in ecosystems–Nature's response to the 'Law of the minimum.' In: Hawksworth DL. Frontiers in mycology. Wallingford (United Kingdom): CAB International. pp. 101–130.
20. Marler MM, Zabinski CA, Callaway RM. (1999) Mycorrhizae indirectly enhance competitive effects of an invasive forb on a native bunchgrass. Ecology 80: 1180–1186.
21. Van der Heijden MGA (2004) Arbuscular mycorrhizal fungi as support systems for seedling establishment in grassland. Ecol Lett 7: 293–303.
22. Nuzzo V (1999) Invasion pattern of the herb garlic mustard (*Alliaria petiolata*) in high quality forests. Biol Invasions 1: 169–179.
23. Nuzzo V (2000) Element stewardship abstract for *Alliaria petiolata*. Arlington (Virginia): The Nature Conservancy. Available: http://tncweeds.ucdavis.edu/esadocs/documnts/allipet.html. Accessed 16 March 2006.
24. Blossey B, Nuzzo V, Hinz H, Gerber E. (2001) Developing biological control of *Alliaria petiolata* (M. Bieb.) Cavara and Grande (garlic mustard). Nat Areas J 21: 357–367.
25. Meekins JF, McCarthy BC (1999) Competitive ability of *Alliaria petiolata* (garlic mustard, Brassicaceae), an invasive, nonindigenous forest herb. Int J Plant Sci 160: 743–752.
26. Prati D, Bossdorf O (2004) Allelopathic inhibition of germination by *Alliaria petiolata* (Brassicaceae). Am J Bot 91: 285–288.
27. Roberts KJ, Anderson RC (2001) Effect of garlic mustard [*Alliaria petiolata* (Beib. Cavara & Grande)] extracts on plants and arbuscular mycorrhizal (AM) fungi. Am Midl Nat 146: 146–152.
28. Renwick JAA (2002) The chemical world of crucivores: Lures, treats and traps. Entomol Exp Appl 104: 35–42.
29. Siemens DS, Garner S, Mitchell-Olds T, Callaway RM (2002) Cost of defense in the context of plant competition: *Brassica rapa* may grow and defend. Ecology 83: 505–517.
30. Schreiner RP, Koide RT (1993) Mustards, mustard oils and mycorrhizas. New Phytol 123: 107–113.
31. Vaughn SF, Berhow MA (1999) Allelochemicals isolated from tissues of the invasive weed garlic mustard (*Alliaria petiolata*). J Chem Ecol 25: 2495–2504.
32. Klironomos JN (2003) Variation in plant response to native and exotic arbuscular mycorrhizal fungi. Ecology 84: 2292–2301.
33. Newsham KK, Fitter AH, Watkinson AR (1995) Multifunctionality and biodiversity in arbuscular mycorrhizas. Trends Ecol Evol 10: 407–411.
34. Reinhart KO, Packer A, Van der Putten WH, Clay K (2003) Plant-soil biota interactions and spatial distribution of black cherry in its native and invasive ranges. Ecol Lett 6: 1046–1050.
35. Callaway RM, Thelen C, Rodriguez A, Holben WE (2004) Soil biota and exotic plant invasion. Nature 427: 731–733.
36. Callaway RM, Thelen GC, Barth S, Ramsey PW, Gannon JE (2004) Soil fungi alter interactions between the invader *Centaurea maculosa* and North American natives. Ecology 85: 1062–1071.
37. Brundrett MC, Piche Y, Peterson RL (1984) A new method for observing the morphology of vesicular-arbuscular mycorrhizae. Can J Bot 62: 2128–2134.
38. McGonigle TP, Miller MH, Evans DG, Fairchild GL, Swan JA (1990) A new method which gives an objective measure of colonization of roots by vesicular arbuscular mycorrhizal fungi. New Phytol 115: 495–501.
39. Klironomos JN, Allen MF, Rillig MC, Piotrowski J, Makvandi-Nejad S, et al. (2005) Abrupt rise in atmospheric CO_2 overestimates community response in a model plant-soil system. Nature 433: 621–624.

Guiding Questions for Reading This Article

A. About the Article

1. Give the name of the journal and the year in which this article was published.

2. What is the last name of the first author, and what is her university?

3. The seven authors represent five different universities in what three countries?

4. Specialized vocabulary: Write a brief definition of each term.

   allelopathy

   arbuscular mycorrhizal fungi (AMF)

   invasive species

   mesic

   mutualism

5. The first paragraph of the article gives several reasons why introduced exotic plants can become invasive and disturb native plants in the area. State two of those reasons.

6. Write out the full genus and species name, as well as the common name, of the invasive plant in this study. This species is native to what part of the world?

7. What is the typical habitat of garlic mustard in its normal home range? Garlic mustard has become an aggressive invader in what habitats in North America?

8. Garlic mustard plants do not associate with mycorrhizal fungi. What relationship had been observed, prior to this study, between growth of garlic mustard and growth potential for AMF in nearby soil?

B. About the Study

9. Experiment 1 is a test of whether the ability of native tree seedlings to form mycorrhizal associations is affected by what?

10. Read the first part of the detailed methods for Experiment 1 ("Materials and Methods," p. 730 of the original article), and state where the investigators obtained soil for the greenhouse pots in which they planted seedlings.

11. In Figure 1A (part of Experiment 1), what is represented on the *x*-axis and on the *y*-axis? What do different shades of the histogram bars represent?

12. In Figure 1A, the heights of the bars and lines represent the *mean* and *standard error* of measures for each sample. What do these two terms measure?

13. Summarize the results for red maple seedlings from Figure 1A in a simple sentence.

14. In Figure 1B, what is represented on the *y*-axis? From the detailed methods (p. 730), tell how long the seedlings were allowed to grow in the test soils before roots and shoots were harvested, dried, and weighed to get the data for Figure 1B. How were the "sterile" soil samples treated differently from the other samples?

15. The detailed methods for Figure 1B say investigators used a "complete 4×3 factorial design." To what "factors" do the 4 and 3 refer? Name *two* experimental conditions (controlled factors) that were held constant for all samples.

16. For the experiments shown in Figure 1B, how many replicates of each treatment combination were used? What are "replicates," and why should replicates be included in the experiment?

17. The investigators examined two alternate hypotheses for how garlic mustard affected tree seedlings:

    Hypothesis A: Chemicals in soils from areas invaded by the garlic mustard plants affect tree seedlings directly, whether or not mycorrhizal microbes (AMFS) are present.

    Hypothesis B: Chemicals in soils from areas invaded by the garlic mustard plants affect the tree seedlings only if mycorrhizal microbes (AMFS) are present.

    Look at the data in Figure 1B to see how seedling growth varies in sterile and nonsterile soils and in invaded and uninvaded soils. From the data in Figure 1B, which hypothesis, A or B, should be *rejected,* and why?

18. In Experiment 3, the investigators tested the effects of chemical extracts from several different plants on the growth of three tree species (sugar maple, red maple, and white ash). What was the control treatment?

19. Examine the results of Experiment 3 in Figures 3A and 3B (p. 730). What were the main effects of the five different extracts (gm, three tree species, control) on the tree seedling colonization by mycorrhizal fungi? How did the five different extracts affect tree seedling growth in biomass?

20. Make a prediction: What if investigators had used extracts from different ash and maple species that

were native to Europe? Would results be similar to the histograms in Figure 3B?

21. In Experiment 4 (Figure 3C, p. 729), investigators collected spores from two common AMF species and grew each on separate agar plates containing the five different types of extracts as in Experiment 3. State the genus names of the two types of AMF that were used in Experiment 4. What was the effect of garlic mustard extract on germination of each AMF spore species, in comparison to spore germination in the control extract? Why was it informative to test the two fungal species separately from each other?

22. Is this an *observational study,* in which quantitative, observational data are taken but no experimental manipulation is made? Or, is this an *experimental study,* in which researchers make manipulations by which the effects of different variables are tested, one at a time?

23. Is this a *field study,* with data collected on organisms in their natural habitat, or is this a *lab study,* in which plants are studied under controlled conditions in the laboratory or greenhouse?

C. General Conclusions and Extensions of the Work

24. What significant, though indirect, impact of an introduced species on native plants is demonstrated in this study?

25. Imagine that you were a member of this research team and involved in these experiments. What could be a possible follow-up test that extends this work? Briefly state another experiment or measurement you would do within this research system.

ARTICLE 6

Inquiry Figure 41.4: *Can Diet Influence the Frequency of Birth Defects?*

Introduction—The Article and Phenomenon Under Study

Birth defects that result from embryonic abnormalities in neural tube development include spina bifida and anencephaly. For decades, researchers have worked to understand possible causes of neural tube defects (NTDs), both genetic and environmental, and to develop health care practices to reduce their incidence. The British physician R. W. Smithells led pioneering work on possible prevention of NTDs by administration of vitamins to mothers around the time of conception. This research, from the following paper, is the subject of Inquiry Figure 41.4 in *Campbell Biology*, Ninth Edition:

R. W. Smithells et al., Possible prevention of neural-tube defects by periconceptional vitamin supplementation, *Lancet* 315: 339–340 (1980).

Read the complete article beginning on the next page and then answer the questions following the article.

▼ Figure 41.4   **INQUIRY**

Can diet influence the frequency of birth defects?

EXPERIMENT Richard Smithells, of the University of Leeds, in England, examined the effect of vitamin supplementation on the risk of neural tube defects. Women who had had one or more babies with such a defect were put into two study groups. The experimental group consisted of those who were planning a pregnancy and began taking a multivitamin at least four weeks before attempting conception. The control group, who were not given vitamins, included women who declined them and women who were already pregnant. The numbers of neural tube defects resulting from the pregnancies were recorded for each group.

RESULTS

Group	Number of infants/fetuses studied	Infants/fetuses with a neural tube defect
Vitamin supplements (experimental group)	141	1 (0.7%)
No vitamin supplements (control group)	204	12 (5.9%)

CONCLUSION This study provided evidence that vitamin supplementation protects against neural tube defects, at least after the first pregnancy. Follow-up trials demonstrated that folic acid alone provided an equivalent protective effect.

SOURCE R. W. Smithells et al., Possible prevention of neural-tube defects by periconceptional vitamin supplementation, *Lancet* 315: 339–340 (1980).

INQUIRY IN ACTION Read and analyze the original paper in *Inquiry in Action: Interpreting Scientific Papers*.

WHAT IF? Subsequent studies were designed to learn if folic acid supplements prevent neural tube defects during first-time pregnancies. To determine the required number of subjects, what type of additional information did the researchers need?

THE LANCET, FEBRUARY 16, 1980

Preliminary Communication

POSSIBLE PREVENTION OF NEURAL-TUBE DEFECTS BY PERICONCEPTIONAL VITAMIN SUPPLEMENTATION

R. W. Smithells   S. Sheppard

C. J. Schorah

Department of Pædiatrics and Child Health, University of Leeds

M. J. Seller

Pædiatric Research Unit, Guy's Hospital, London

N. C. Nevin

Department of Medical Genetics, Queen's University of Belfast

R. Harris   A. P. Read

Department of Medical Genetics, University of Manchester

D. W. Fielding

Department of Pædiatrics, Chester Hospitals

Summary   Women who had previously given birth to one or more infants with a neural-tube defect (NTD) were recruited into a trial of periconceptional multivitamin supplementation. 1 of 178 infants/fetuses of fully supplemented mothers (0·6%) had an NTD, compared with 13 of 260 infants/fetuses of unsupplemented mothers (5·0%).

INTRODUCTION

The well-known social-class gradient in the incidence of neural-tube defects (NTD) suggests that nutritional factors might be involved in NTD ætiology. A possible link between folate deficiency and NTDs in man was first reported in 1965.[1] More recently, significant social-class differences in dietary intakes in the first trimester,[2] and in first-trimester values for red cell folate, leucocyte ascorbic acid, red-blood-cell riboflavin, and serum vitamin A have been reported,[3] dietary and biochemical values being higher in classes I and II than in classes III, IV, and V. Furthermore, 7 mothers, of whom 6 subsequently gave birth to NTD infants and 1 to an infant with unexplained microcephaly, had first-trimester mean values for red cell folate and leucocyte ascorbic acid that were significantly lower than those of controls.[3]

These observations are compatible with the hypothesis that subclinical deficiencies of one or more vitamins contribute to the causation of NTDs. We report preliminary results of an intervention study in which mothers at increased risk of having NTD infants were offered periconceptional multivitamin supplements.

PATIENTS AND METHODS

Women who had had one or more NTD infants, were planning a further pregnancy, but were not yet pregnant were admitted to the study. All women referred to the departments involved in the study and who met these criteria were invited to take part. Most patients were recruited from genetic counselling clinics, although some were referred by obstetricians and general practitioners informed of the study. Patients came from Northern Ireland, South-East England, Yorkshire, Lancashire, and Cheshire. 185 women who received full vitamin supplementation (see below) became pregnant.

The control group comprised women who had had one or more previous NTD infants but were either pregnant when referred to the study centres or declined to take part in the study. Some centres were able to select a control for each supplemented mother, matched for the number of previous NTD births, the estimated date of conception, and, where possible, age. There were 264 control mothers. The numbers of fully supplemented (S) and control (C) mothers in each centre were as follows: Northern Ireland S 37, C 122; South-East England S 70, C 70; Yorkshire S 38, C 35; Lancashire S 31, C 27; Cheshire S 9, C 10.

All mothers in supplemented and control groups were offered amniocentesis. 6 mothers in Northern Ireland (3 supplemented; 3 controls) declined amniocentesis and their pregnancies continue. They are not included in the figures above or in the accompanying table. All mothers with raised amniotic-fluid alpha-fetoprotein (AFP) values (1 supplemented; 11 controls) accepted termination of pregnancy.

Study mothers were given a multivitamin and iron preparation ('Pregnavite Forte F' Bencard), 1 tablet three times a day for not less than 28 days before conception and continuing at least until the date of the second missed period—i.e., until well after the time of neural-tube closure. Pregnavite forte F provides daily vitamin A 4000 I.U., vitamin D 400 I.U., thiamine 1·5 mg, riboflavin 1·5 mg, pyridoxine 1 mg, nicotinamide 15 mg, ascorbic acid 40 mg, folic acid 0·36 mg, ferrous sulphate equivalent to 75·6 mg Fe, and calcium phosphate 480 mg. Women conceiving less than 28 days after starting supplementation, or starting supplementation shortly after conception, or known to have missed tablets for more than 1 day, are regarded as partly supplemented. They were excluded from the main study and their results will be considered elsewhere.

In Northern Ireland, Yorkshire, and Cheshire women taking oral contraceptives (OCs) were asked to adopt alternative means of contraception from the date of starting vitamins because OCs may lower blood levels of certain vitamins.[4]

RESULTS

187 control mothers have delivered 192 infants (including 5 twin pairs) without NTDs, and a further 38 have normal amniotic-fluid AFP values (table). 13 mothers have been delivered of NTD infants/fetuses, 1

OUTCOME OF PREGNANCY IN FULLY SUPPLEMENTED AND CONTROL MOTHERS

	Fully supplemented	Controls
Infant/fetus with NTD	1	12
Infant without NTD	140(3)	192(5)
Subtotal (1)	141(3)	204(5)
Normal amniotic AFP	26	38
Subtotal (2)	167(3)	242(5)
Spontaneous abortions		
Examined, NTD	0	1
Examined, no NTD	11	17
Subtotal (3)	178(3)	260(5)
Not examined	10	9
Total	188(3)	269(5)

All numbers relate to infants/fetuses.
Figures in parentheses indicate numbers of twin pairs included.

THE LANCET, FEBRUARY 16, 1980

by spontaneous abortion, 11 by termination after amniocentesis, and 1 by spontaneous delivery (skin-covered lesion, normal AFP). 17 fetuses of a further 26 control mothers who aborted spontaneously were examined and had no NTD. The provisional recurrence-rate of NTDs is 5·0% (13 in 260). 26 control mothers were at increased risk by virtue of having had 2 previous NTD infants. 3 of them had a further affected child, a recurrence-rate of 11·5%. Both these recurrence-rates are consistent with those previously reported and widely adopted in genetic counselling.

137 fully supplemented mothers have given birth to 140 babies (including 3 twin pairs) without NTD, 26 have normal amniotic-fluid AFP values and their pregnancies continue, and 1 has had a further affected infant. 11 fetuses of 21 mothers who aborted spontaneously were examined; none had an NTD. The provisional recurrence-rate in the supplemented group is therefore 0·6% (1 in 178). 15 supplemented mothers were at increased risk by virtue of having had 2 previous affected NTD infants. None had a further affected child.

Comparison of NTD frequencies in the supplemented and control groups by Fisher's exact test showed significant differences (p<0·01) for subtotals (1), (2), and (3) (table).

DISCUSSION

Despite problems with choosing controls, the control women in this study have shown recurrence-rates for NTDs entirely consistent with published data. By contrast the supplemented mothers had a significantly lower recurrence-rate. Possible interpretations of this observation include the following:

(1) *A group of women with a naturally low recurrence risk has unwittingly selected itself for supplementation.*—Apart from geographic and secular variations there is no evidence to suggest that any particular subgroup within populations, whether by social class or any other division, has a higher or lower recurrence risk. In genetic counselling clinics it is customary to quote the same risk for all mothers after one affected child. We cannot exclude the possibility that women who volunteered and cooperated in the trial might have had a reduced risk of recurrence of NTD. However, one might have expected such an effect to be found in mothers who cooperated in potato-avoidance trials, but this was not seen.[5]

(2) *Supplemented mothers aborted more NTD fetuses than did controls.*—The proportion of pregnancies ending in spontaneous abortion is similar in the two groups (supplemented 11·4%, control 9·6%). If the supplemented mothers have aborted more NTD fetuses, they must have aborted fewer other fetuses or had a lower initial risk of abortion. 11 of 21 abortuses of supplemented mothers have been examined and none had an NTD. 18 of 27 abortuses of control mothers were examined and 1 had an NTD. An explanation based on selective abortion of fetuses with NTD seems improbable, especially since more abortions are likely to have been ascertained in the supplemented group since controls were enrolled later in pregnancy.

(3) *Something other than vitamin supplementation has reduced the incidence of NTDs in the treated group.*—This is an almost untestable hypothesis, but if anything has reduced the incidence of NTDs it needs to be identi-

fied urgently. The only measure introduced by the study other than vitamin supplementation (and that only in some centres) was discontinuation of OCs at least 28 days before conception. Although the possibility of sex hormones having teratogenic action is not yet entirely resolved, evidence[6] strongly suggests that the phenomenon we report is not attributable to stopping OCs.

(4) *Vitamin supplementation has prevented some NTD.*—This is the most straightforward interpretation and is consistent with the circumstantial evidence linking nutrition with NTDs. If the vitamin tablets are directly responsible, we cannot tell from this study whether they operate via a nutritional or a placebo effect.

We hope that the data presented will encourage others to initiate similar and related studies. We intend to publish a more detailed report when the last of the present cohort of women receiving vitamin supplements has had her baby (due April 1980).

We thank the women taking part in this study; medical colleagues who referred them; and Dr Jennifer Hanna; Miss Wendy Johnston, Mrs Monica Stant and Mrs Mary Weetman (health visitors). This study is supported by Action Research for the Crippled Child, the Children's Research Fund, and Beecham Pharmaceuticals Ltd.

Requests for reprints should be addressed to R. W. S., Department of Pædiatrics and Child Health, University of Leeds, 27 Blundell Street, Leeds LS1 3ET.

REFERENCES

1. Hibbard ED, Smithells RW. Folic acid metabolism and human embryopathy. *Lancet* 1965; i: 1254–56.
2. Smithells RW, Ankers C, Carver ME, Lennon D, Schorah CJ, Sheppard S. Maternal nutrition in early pregnancy. *Br J Nutr* 1977; **38**: 497–506.
3. Smithells RW, Sheppard S, Schorah CJ. Vitamin deficiencies and neural tube defects. *Arch Dis Childh* 1976, **51**: 944–50.
4. Wynn V. Vitamins and oral contraceptive use. *Lancet* 1975; i: 561–64.
5. Nevin NC, Merrett JD. Potato avoidance during pregnancy in women with a previous infant with either anencephaly and/or spina bifida. *Br J Prev Soc Med* 1975; **29**: 111–15.
6. Rothman KJ, Louik C. Oral contraceptives and birth defects. *N Engl J Med* 1978; **299**: 522–24.

Guiding Questions for Reading This Article

A. About the Article

1. Give the name of the journal and the year in which this article was published.

2. State the last name of the first author, his department, and his university.

3. Specialized vocabulary: Write a brief definition of each term.

   amniocentesis

   neural tube defects (NTDs)

   periconceptional

   placebo

   vitamin

B. About the Study

4. The authors point out that the observed higher incidence of NTDs in lower social classes as compared to higher social classes might be due to what factors?

5. What criteria were used to select women for this study?

6. The control group consisted of whom?

7. How many mothers were in the fully supplemented group, and what was their treatment?

8. How did investigators treat data on women who conceived before taking the supplements for a month and those who missed some of the supplements?

9. In controlled experiments in general, the experimental group and the control group are alike in all factors except in the one being tested. In this study, the test factor is the nutritional supplement. (a) Some study centers used paired controls, in which a supplemented mother was paired for comparison with a control mother. What criteria were used for matching the pairs? (b) In this study, what are some other ways the supplemented mothers and the control mothers might have differed, besides whether or not they received the supplement?

10. From the table showing outcome of pregnancy in fully supplemented and control mothers, what is the difference in number of NTD infants between the supplemented and control groups? What was the difference in percentage of NTDs in the two groups of women?

11. The authors state that their data agree with the hypothesis that vitamin supplementation during the period around conception is associated with lower incidence of NTDs. In their Discussion section, they mention three alternative explanations for this association. Briefly list the three explanations in your own words.

C. General Conclusions and Extensions of the Work

12. Do you think the observed difference is significant enough to conclude that vitamin supplementation has prevented some NTDs in women who have previously had NTD infants? Do you think the results can be generalized to conclude that vitamin supplementation will prevent NTDs in all women? Why or why not?

13. In 1983, B. Lipsett and J. C. Fletcher published a paper entitled "Do vitamins prevent neural tube defects (and can we find out ethically)?" in the *Hastings Center Report* (13:508). They documented the early history of R. W. Smithells's work on multivitamins and birth defects, including the paper in this exercise. They pointed out that, before beginning his studies in 1976, Smithells had requested approval from several ethics committees to do a "randomized, placebo-controlled" clinical trial, but his requests were refused. (a) How would the study procedures be different if trials were "randomized"? (b) What is a "placebo"? How would the study procedures be different with the use of placebos? (c) Why do you think the ethics committees denied Smithells's request? Do you think they should have approved the research request?

14. Imagine that you were a member of this research team and involved in these investigations. What could be a possible follow-up test that extends this work?

Inquiry Figure 56.13: *What Caused the Drastic Decline of the Illinois Greater Prairie Chicken Population?*

Introduction—The Article and Phenomenon Under Study

As populations become threatened because of habitat loss, both small population size and low genetic variability can contribute to their decline and lead to an "extinction vortex." Both factors must be considered in deciding upon appropriate conservation strategies, but it is unusual to have sufficient long-term data on endangered populations. A 35-year history of greater prairie chicken populations, including the results of a conservation intervention, provides a rare opportunity to study such an endangered species. This study, from the following paper, is the subject of Inquiry Figure 56.13 in *Campbell Biology,* Ninth Edition:

R. L. Westemeier et al., Tracking the long-term decline and recovery of an isolated population, *Science* 282:1695–1698 (1998).

Read the complete article beginning on the next page and then answer the questions following the article.

▼ Figure 56.13 **INQUIRY**

What caused the drastic decline of the Illinois greater prairie chicken population?

EXPERIMENT Researchers had observed that the population collapse of the greater prairie chicken was mirrored in a reduction in fertility, as measured by the hatching rate of eggs. Comparison of DNA samples from the Jasper County, Illinois, population with DNA from feathers in museum specimens showed that genetic variation had declined in the study population (see Figure 23.11). In 1992, Ronald Westemeier, Jeffrey Brawn, and colleagues began translocating prairie chickens from Minnesota, Kansas, and Nebraska in an attempt to increase genetic variation.

RESULTS After translocation (blue arrow), the viability of eggs rapidly increased, and the population rebounded.

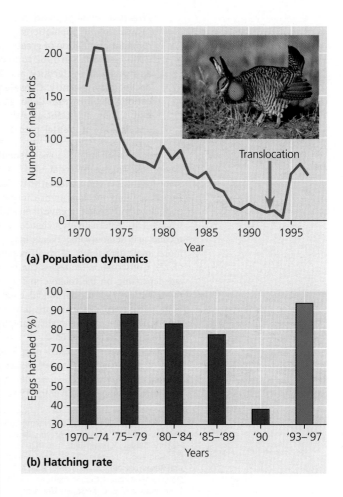

(a) Population dynamics

(b) Hatching rate

CONCLUSION Reduced genetic variation had started the Jasper County population of prairie chickens down the extinction vortex.

SOURCE R. L. Westemeier et al., Tracking the long-term decline and recovery of an isolated population, *Science* 282:1695–1698 (1998).

INQUIRY IN ACTION Read and analyze the original paper in *Inquiry in Action: Interpreting Scientific Papers.*

WHAT IF? Given the success of using transplanted birds as a tool for increasing the percentage of hatched eggs in Illinois, why wouldn't you transplant *additional* birds immediately to Illinois?

Tracking the Long-Term Decline and Recovery of an Isolated Population

Ronald L. Westemeier,* Jeffrey D. Brawn,† Scott A. Simpson,
Terry L. Esker, Roger W. Jansen, Jeffery W. Walk,
Eric L. Kershner, Juan L. Bouzat, Ken N. Paige

Effects of small population size and reduced genetic variation on the viability of wild animal populations remain controversial. During a 35-year study of a remnant population of greater prairie chickens, population size decreased from 2000 individuals in 1962 to fewer than 50 by 1994. Concurrently, both fitness, as measured by fertility and hatching rates of eggs, and genetic diversity declined significantly. Conservation measures initiated in 1992 with translocations of birds from large, genetically diverse populations restored egg viability. Thus, sufficient genetic resources appear to be critical for maintaining populations of greater prairie chickens.

The conservation implications of small population size are controversial (1–4). A significant loss in genetic variation may decrease fitness or limit the long-term capacity of a population to respond to environmental challenges (5). Alternatively, chance environmental and demographic events may pose a more immediate threat to small populations (1, 2). Conservation strategies can be different depending on the relative importance of these factors (1, 3, 6), but fundamental questions persist because there are few data on long-term changes in the demography and genetics of wild populations.

Here we report the results of a long-term study on a remnant population of greater prairie chickens (*Tympanuchus cupido pinnatus*) in southeastern Illinois (7). Over the 35-year peri-od of this study, we documented concurrent declines in population size and fitness as well as an overall reduction in genetic diversity. In addition, we report on a conservation strategy initiated in 1992, whereby translocations of individuals from large, genetically diverse populations increased fitness.

Greater prairie chickens are grassland-dependent birds still found in areas of suitable habitat ranging from northwestern Minnesota south to northeastern Oklahoma, and from southeastern Illinois west to northeastern Colorado (8). Leks (or booming grounds) are used as arenas for territorial display and breeding by two or more males (9). Loss of habitat suitable for successful nesting and brood rearing is the single most important factor leading to declines, isolation, and extirpations throughout the species' range in the midwestern United States (10). The eastern subspecies *Tympanuchus cupido cupido*, also known as the heath hen, has been extinct since 1931 (11) and Attwater's prairie chicken *Tympanuchus cupido attwateri*, which is restricted to Texas, is near extinction (12, 13).

In Illinois, native prairie habitat for prairie chickens originally covered >60% of the state (Fig. 1), but fewer than 931 ha (<0.01%) of the original 8.5×10^6 ha of high-grade prairie remain (14). There were possibly several million prairie chickens statewide in the mid-19th century (15); by 1962 an estimated 2000 birds

R. L. Westemeier, Illinois Natural History Survey, Effingham, IL 62401, USA. J. D. Brawn, Illinois Natural History Survey, Champaign, IL 61820, USA. S. A. Simpson and T. L. Esker, Illinois Department of Natural Resources, Newton, IL 62448, USA. R. W. Jansen, Douglas-Hart Nature Center, Mattoon, IL 61938, USA. J. W. Walk and E. L. Kershner, University of Illinois, Department of Natural Resources and Environmental Sciences, Urbana, IL 61801, USA. J. L. Bouzat and K. N. Paige, University of Illinois, Department of Ecology, Ethology, and Evolution, Urbana, IL 61801, USA.

*To whom correspondence should be addressed.
†To whom e-mail should be addressed at j-brawn@uiuc.edu

were reported in 179 localized groups occupying about 1500 km² in 15 counties in southern Illinois(*16*). Although early efforts (1963–1973) to preserve prairie chickens in Illinois showed marked success on restored grassland habitat (*17*), by 1994 an estimated 46 birds remained on about 33 km² in two small populations (*18–20*). These remnant populations were geographically isolated from larger, more contiguous populations about 640 km to the west (*8*). Estimated size of the focal population in Jasper County, Illinois, fluctuated from 84 males in 1963 to about 40 in the mid-1960s and then increased markedly to a high of 206 males counted on 13 leks in 1972 (*21*). By spring 1994, only five or six Illinois males remained on one unstable lek (Fig. 2). The drop to near extirpation occurred despite an increase in local availability of managed grassland habitat be-

tween 1963 and 1994 (*22*).

Fertility (fertile incubated eggs per total eggs) and success (hatched eggs per total eggs in fully incubated clutches) rates of eggs are key fitness traits in birds (*23*). These traits also fluctuated over time but decreased significantly in the focal population between 1963 and 1991 (Fig. 2) (*24*). The overall fertility rate (based on 3357 eggs) for 278 clutches was 93% and was sustained at >90% through 1980. Fertility rates declined in the subsequent 12 years with a low of 74% in 1990. In the 1960s, rates of egg success ranged from 91% to 100%, but by 1981 and in all but three of the next 10 breeding seasons, success rates lower than 80% were observed. The decreasing trend was significant even without the extremely low egg success of 38% observed in 1990 (ϕ = 3.35; P < 0.001) (Fig. 2). About 50% of the nests observed

before 1981 had partial hatching failure, but only 10% had four or more eggs failing. After 1980, 70% of successful nests contained at least one unhatched egg and failure of four or more eggs per nest was increasingly common (43%). Even two fewer chicks per brood may decrease recruitment rates of prairie chickens (*13*).

Rates of egg success observed in the focal population after 1980 were markedly lower than the 93% rate observed in the same county during the 1930s (*25*), when statewide abundance was estimated at 25,000 birds (*26*). Other data from larger prairie chicken populations to the west, northwest, and north, many during the years covered by our study, reveal an overall average egg success rate of 94% (*n* = 216 successful clutches) with a range among 18 studies of 80% to100% (*25, 27, 28*). Clearly, egg success experienced by the isolated Illinois population after 1980 was unusually low for the species (*29*).

Estimated genetic variation within Illinois' focal population was markedly lower than that within samples from larger populations in Kansas, Nebraska, and Minnesota (*30*). When sampled (1992–1994), all three of these populations numbered in the thousands, but Illinois' focal population probably did not exceed 250 birds when sampled during 1974–1993. The Illinois population had the lowest estimated mean heterozygosity and about two-thirds the allelic diversity observed in the larger populations (*31*). All alleles detected in the Illinois population (for 1974–1993) were present in one or more of the larger populations. Several alleles common to the large populations were not detected in the Illinois population but were known to be present before the 1970s (*30*).

The poor reproductive performance and inevitable extirpation of the Illinois population led local managers to initiate a translocation program in August 1992. Objectives were to increase numbers and enhance the genetic diversity and fitness within the focal population. Between 1992 and 1996, 271 greater prairie chickens (144 females and 127 males) were transplanted from large populations in Minnesota, Kansas, and Nebraska. Radiotelemetry data and observations of banded birds on leks indicated that, after each release, 25% to 67% of the transplanted prairie chickens survived and integrated per year into the breeding population (*20*). Although four radio telemetry–tagged hens from Minnesota nested in 1993, recruitment of young was not verified until 1994, when the mixed population of Illinois, Minnesota, and Kansas birds bred. From the low count of five or six Illinois males (plus two Minnesota males) in 1994 on one unstable lek, the spring count in 1996 had increased to 70 males of mixed origin on six leks (Fig. 2).

Eggs in 14 successful nests located in 1993, 1994, and 1997 revealed significant increases from the previous decade (1982–1991) in mean rates of fertility (91% to 99%; Mann-Whitney tests, Z = 2.32, P < 0.05) and hatching (76% to

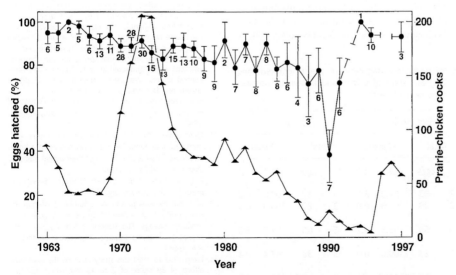

Fig. 1. Illinois prairies during 1810–1820 and distributions of greater prairie chickens in 1940 (*25*), 1962, and 1994. Prairie distributions for 1810–1820 were derived from R. C. Anderson. [Reprinted from R. C. Anderson, *Transactions of the Illinois State Academy of Science* **63**, 214 (1970), with permission.]

Fig. 2. Annual means for success of greater prairie chicken eggs in 304 fully incubated clutches (circles) and counts of males (triangles) on booming grounds in spring, Jasper County, Illinois, 1963–1997. Translocations of nonresident birds began in August 1992. Test statistics (*24*) for the period 1963–1991 are as follows: egg success rates, ϕ = 4.28 (P < 0.001); male counts, ϕ = 1.88 (P = 0.0301). Bars indicate ±1 SE and adjacent numbers indicate numbers of nests. For egg fertility rates (not shown), ϕ = 2.18 (P = 0.0146).

REPORTS

94%; $Z = 2.80$, $P < 0.01$) (Fig. 2). Identities were known for eight of the hens associated with these 14 nests because they were radio telemetry–tagged transplanted birds (Table 1). During 1993 and 1994, three hatches involved two Minnesota hens, five hatches involved Kansas hens, and three hatches were by hens of unknown origin; males could have been of Illinois (one definite in 1993), Minnesota, or Kansas origin (20). In spring 1994, five Illinois males and two Minnesota males were in situ territory holders and may have predominated in matings with Minnesota hens and newly introduced hens from Kansas. We lack nest data for 1995 and 1996; the three successful clutches found in 1997 were likely derived from birds of mixed origin. Unfortunately, we knew of no successful clutches from Illinois × Illinois crosses that might have served as controls during the 1993, 1994, and 1997 breeding seasons.

We know of no major environmental or climatic changes from previous years on the focal population's breeding grounds that could account for increased egg viability in 1993, 1994, and 1997. In Illinois, May is a critical month for incubation of most prairie chicken clutches and for hatching of the earliest clutches. Excessive rainfall and flooding, or too little rain, have been correlated with decreased reproductive success of prairie chickens (12). In 1963–1991, egg success correlated inversely with precipitation in May ($r = -0.36$; $n = 29$; $P < 0.05$). In May 1993, 1994, and 1997, respective rainfall amounts were 85%, 49%, and 91% of normal (32), and rates of egg success for those years were the highest in 25 years. Rainfall amounts were similarly favorable during much of the 1970s and 1980s when egg success was declining. Other factors (33) that might have suppressed egg fertility and success were still present on the breeding grounds when egg viability was restored. Therefore, egg success likely increased after the transplant because of factors intrinsic to the breeding birds.

Greater prairie chickens in Illinois illustrate the challenge of conserving small populations. From 1962 to the present, intensive management was carried out with the goal of increasing population size. Marked increases occurred from 1968 to 1972. The quality and quantity of managed habitat were enhanced and both nest parasites and predators were controlled (22). Yet, despite these efforts and successes, overall population size and fitness decreased. A key demographic event in the 1970s may have doomed the viability of the focal population. Before that time, the population was surrounded by other, albeit smaller, populations within 8 km; during the 1960s gene flow among those populations was likely (Fig. 1). The satellite populations subsequently disappeared and, by 1980, nearly all breeding by greater prairie chickens in the region was on or within 0.8 km of managed grasslands. Once isolated, the focal population lost viability and a conservation strategy designed to enhance genetic variation became necessary.

The demographic performance of the focal population is also a unique example of the general scenario predicted by "extinction vortex" models (34). These models predict that demographic and genetic effects reinforce each other in small populations to increase the probability of local extinction. We believe that the near-complete loss of suitable grasslands and satellite populations in the region drove the greater prairie chicken toward this scenario. Small population size and isolation then led to low genetic diversity and decreased fitness. The declining numbers and fitness were not unlike those of the now extinct heath hen (34). Attwater's prairie chicken also showed a similar decline in numbers, with egg success of 93% in 1937 but as low as 50% by 1985 (13), and two of the three remaining populations showed reduced genetic variability (35). Without intervention, our focal population likely would not have recovered genetic variation sufficient to offset adverse effects on vital demographic traits. We predict that periodic translocations will be necessary to maintain the focal population unless significantly more habitat becomes available. Isolated relict populations, such as greater prairie chickens in Illinois, cannot be conserved indefinitely with inadequate habitat and small size.

References and Notes
1. R. Lande, *Science* **241**, 1455 (1988).
2. T. M. Caro and M. K. Laurenson, *ibid.* **263**, 485 (1994).
3. G. Caughley, *J. Anim. Ecol.* **63**, 215 (1994).
4. R. Lande, *Conserv. Biol.* **9**, 782 (1995).
5. L. F. Keller, P. Arcese, J. N. M. Smith, W. M. Hochachka, S. C. Stearns, *Nature* **372**, 356 (1994); R. Frankham and K. Ralls, *ibid.* **392**, 441 (1998); I. Saccheri *et al.*, *ibid.*, p. 491.
6. P. W. Hedrick, R. C. Lacy, F. W. Allendorf, M. E. Soule, *Conserv. Biol.* **10**, 1312 (1996).
7. R. L. Westemeier, J. E. Buhnerkempe, J. D. Brawn, *Wilson Bull.* **110**, 190 (1998).
8. P. A. Johnsgard, *The Grouse of the World* (Univ. of Nebraska Press, Lincoln, NE, 1983), pp. 316–340.
9. F. and F. Hamerstrom, *Technical Bulletin No. 64* (Department of Natural Resources, Madison, WI, 1973).
10. M. A. Wisdom and L. S. Mills, *J. Wildl. Manage.* **61**, 302 (1997).
11. A. O. Gross, *Mass. Audubon Soc. Bull.* **15**, 12 (1932).
12. M. E. Morrow, R. S. Adamcik, J. D. Friday, L. B. McKinney, *Wildl. Soc. Bull.* **24**, 593 (1996).
13. M. J. Peterson and N. J. Silvy, *Conserv. Biol.* **10**, 1264 (1996).
14. J. White, *Illinois Natural Areas Inventory Technical Report. Vol. 1: Survey Methods and Results* (Illinois Natural Areas Inventory, Urbana, IL, 1978).
15. Illinois State Historical Society, *Fourth Annual Illinois History Symposium*, Springfield, IL, 2–3 December 1983 (Illinois State Historical Society, Springfield, IL, 1985).
16. G. C. Sanderson and W. R. Edwards, *Trans. Ill. State Acad. Sci.* **59**, 326 (1966).
17. *The Prairie Chicken in Minnesota*, University of Minnesota, Crookston, MN, 28 April 1973 (University of Minnesota, Crookston, MN, 1973).
18. R. L. Westemeier, S. A. Simpson, D. A. Cooper, *Wilson Bull.* **103**, 717 (1991).
19. Our field study focused on Jasper County, near Newton, Illinois, where grasslands were restored on scattered sanctuaries developed by The Prairie Chicken Foundation of Illinois, The Nature Conservancy, and the Illinois Department of Natural Resources. Studies of lesser scope were conducted near Kinmundy in Marion County and annual spring surveys of prairie chickens were conducted in up to 23 areas of south-central Illinois. We used standard methods (9) to census prairie chickens during 1963–1997.
20. R. L. Westemeier and R. W. Jansen, *Illinois Natural History Survey Reports No. 332* (Champaign, IL, 1995).
21. Hens observed on leks are not used for annual estimates of abundance because they represent a lesser and more variable proportion of their actual numbers than is typical for males (9).
22. R. L. Westemeier, *Illinois Natural History Survey Reports No. 343* (Champaign, IL, 1997).
23. A. J. van Noordwijk and W. Scharloo, *Evolution* **35**, 674 (1981); F. W. Allendorf and R. F. Leary, in *Conservation Biology, The Science of Scarcity and Diversity*, M. E. Soule, Ed. (Sinauer Associates, Sunderland, MA, 1986).
24. From 1963 to 1991 this study involved the systematic efforts of 29 teams of 2 to 15 researchers and field assistants who searched on foot a cumulative total of nearly 4050 ha for nests of grassland birds; prairie chickens were the key species. Of 1125 prairie chicken nests examined, 1003 were from the focal population in Jasper County. A smaller sample of 99 nests was found in and near a study area in Marion County, 64 km of the primary study area. In 1993, 1994, and 1997, 23 additional nests

Table 1. Clutch size, egg fertility, and egg success for nests involving resident prairie chickens and nonresidents translocated to Jasper County, Illinois, by year, state of origin, and known or possible parentage combinations in 1993, 1994, and 1997. N = number of nests.

Year	State of origin		Clutch size			Egg fertility				Egg success			
	Hens	Cocks	N	Mean	SE	Total		Mean		Total		Mean	
						N	Eggs	%	SE	N	Eggs	%	SE
1993*	MN	IL	3	17.0	0.6	2	33	97.0	3.1	1	17	100.0	0.0
1994*	MN	IL, MN, or KS	2	14.5	0.5	2	29	100.0	0.0	2	29	100.0	0.0
1994*	KS	IL, MN, or KS	5	13.0	1.0	5	64	98.5	1.7	5	65	87.7	5.5
1994†	IL, MN, or KS	IL, MN, or KS	3	12.7	1.0	2	24	100.0	0.0	3	38	97.4	2.4
1997†	IL, MN, KS, or NE	IL, MN, KS, or NE	3	14.3	0.7	2	28	100.0	0.0	3	43	93.3	6.7
All			16	14.1	0.5	13	179	98.9	0.8	14	192	93.8	3.0

*Origin of hens verified by radio telemetry tagging, with individual frequencies.   †Nests found indicidentally; origin of hens and cocks unknown.

REPORTS

were found in Jasper County, 17 of which involved radio telemetry–tagged hens from Minnesota and Kansas. To determine clutch size, fertility, and egg success, we used only nests with eggs or egg shells in good condition at the time of discovery and not those known to be partially or fully depredated. Searches were timed so that about 90% of nests were hatched, depredated, or abandoned upon discovery; about 10% were still active when found. Although researcher disturbance of 47 active nests was suspect in biasing egg success rates, a separate study of this possibility was inconclusive (7). Hence, we tested rates of egg success with and without the disturbed clutches. When only undisturbed nests were used the test statistic was still highly significant (ϕ = 2.95; P = 0.0016). Fertility and egg success were calculated by dividing the number of fertile (germinal discs or embryos evident but eggs not always hatched or fully incubated) or hatched eggs by the total number of eggs in unparasitized, fully incubated clutches. We excluded 38 successful nests parasitized by ring-necked pheasants (*Phasianus colchicus*) in calculating egg success to avoid bias of low success in parasitized nests [R. L. Westemeier, J. E. Buhnerkempe, W. R. Edwards, J. D. Brawn, S. A. Simpson, *J. Wildl. Manage.* **62**, 854 (1998)]. All tests (1963–1991 data) in Fig. 2 and for egg fertility were based on a nonparametric test for trend developed by E. L. Lehmann [*Nonparametrics— Statistical Methods Based on Ranks* (McGraw-Hill, New York, 1975)] using the normal approximation in all cases.

25. R. E. Yeatter, *Ill. Nat. Hist. Surv. Bull.* **22**, 377 (1943).
26. J. Lockart, *Ill. Dep. Conserv. Tech. Bull.* (1968).
27. A. O. Gross, *Wis. Conserv. Comm. Bull.* (1930); F. N. Hamerstrom Jr., *Wilson Bull.* **51**, 105 (1939); C. W. Schwartz, *Univ. Mo. Stud.* **20**, 1 (1945); W. B. Grange, *Wis. Conserv. Dep. Publ.* **328** (1948); M. F. Baker, *Univ. Kans. Mus. Nat. Hist. Biol. Surv. Kans.* **5** (1953); F. L. Arthaud, thesis, University of Missouri (1968); N. J. Silvy, thesis, Kansas State University (1968); R. D. Drobney, thesis, University of Missouri (1973); L. H. Sisson, *The Sharp-Tailed Grouse in Nebraska, a Research Study* (Nebraska Game and Parks Commission, Lincoln, 1976); W. D. Svedarsky, dissertation, University of North Dakota (1979); L. A. Rice and A. V. Carter, *Completion Report No. 84-11* (South Dakota Department of Game, Fish, and Parks, Pierre, SD, 1982); G. J. Horak, *Kansas Fish and Game Commission Wildlife Bulletin No. 3* (1985); J. A. Newell, thesis, Montana State University (1987); D. P. Jones, thesis, University of Missouri (1988); J. E. Toepfer, dissertation, Montana State University (1988).
28. L. L. McDaniel, unpublished material.
29. Our nest data were highly representative of the focal population because, on average during 1963–1991, total nests found represented 46% (range, 28% to 76%) and 77% (range, 39% to 188%) of the number of males and hens, respectively, censused on booming grounds in Jasper County.
30. J. L. Bouzat, H. A. Lewin, K. N. Paige, *Am. Nat.* **152**, 1 (1998); J. L. Bouzat *et al.*, *Conserv. Biol.* **10**, 836 (1998).
31. Estimates of genetic variability were based on highly polymorphic type II markers (microsatellites), which are more sensitive for detecting potential changes in genetic diversity. The relatively small reduction in Illinois average heterozygosity (about 9%) results from the large numbers of alleles detected at the microsatellite loci. Estimates of genetic diversity based on type I markers (that is, markers associated with coding sequences of structural genes) would likely reveal a more drastic decline in average heterozygosity. The observed declines of fitness components in the Illinois population should not be attributed to the particular loci analyzed (which are noncoding and presumably neutral DNA sequences). The effects of inbreeding are probably more directly related to the expression of lethal recessive alleles at structural genes. Our results also indicate that allelic diversity is more drastically affected than heterozygosity, as is expected following population bottlenecks (M. Nei, T. Maruyama, R. Chakraborty, *Evolution* **29**, 1 1975). Declines in fitness in the Illinois population probably result from a drastic decrease in heterozygosity at structural genes and the random extinction of critical alleles through the population bottleneck.
32. Illinois State Water Survey, *Ill. Agric. Stat. Serv.* (1993); *ibid.* (1994); *ibid.* (1997).
33. Factors other than inbreeding can cause fertility and hatchability problems. Other possible factors included competing species that may transmit disease (pheasants and waterfowl), pollutants (oil and pesticides), and human disturbances. However, egg success was declining about one decade before there was a large abundance of pheasants in the mid-1980s, researcher disturbance of active nests (mostly the mid-1980s to 1991) (7), a large increase in oil production (1983), and pesticides used for no-till farming. Similarly, waterfowl, mostly mallards (*Anas platyrhynchos*), did not feed or nest with prairie chickens until the late 1970s. Also, hatch rates of northern bobwhites (*Colinus virginianus*), pheasants, and other sympatric species in the study area have not declined.
34. M. E. Gilpin and M. E. Soule, in *Conservation Biology: The Science of Scarcity and Diversity*, M. E. Soule, Ed. (Sinauer Associates, Sunderland, MA, 1986), pp. 19–34.
35. A. O. Gross, *Mem. Boston Soc. Nat. Hist.* **6**, 491 (1928).
36. E. A. Osterndorff, thesis, Texas A&M University (1995).
37. We thank the many field assistants who made this report possible and especially project assistants D. R. Vance and J. E. Buhnerkempe; R. J. Ellis led the project during 1963–1965. We appreciate technical reviews of the manuscript by K. M. Giesen, C. A. Phillips, and P. W. Brown and editing and figure graphics by T. E. Rice. J. E. Toepfer provided guidance in all phases of translocation efforts and, with P. S. Beringer, helped in securing birds from Minnesota for translocation to Illinois. Illinois wild turkeys were traded for Minnesota prairie chickens with the assistance of the Illinois Department of Natural Resources–Division of Wildlife Resources. Various other wildlife staff in Minnesota, North Dakota, Kansas, and Nebraska supported translocation of prairie chickens. L. L. McDaniel provided unpublished egg success data from the Valentine National Wildlife Refuge, Nebraska. This is a contribution in part of Federal Aid Wildlife Restoration Project W-66-R, with the Illinois Department of Natural Resources–Division of Natural Heritage, U. S. Fish & Wildlife Service, Illinois Natural History Survey, The Nature Conservancy, Illinois Nature Preserves Commission, Illinois Endangered Species Protection Board, and the University of Illinois at Urbana–Champaign; cooperating. Funding for translocations of prairie chickens was provided by the Illinois Wildlife Preservation Fund and Natural Areas Acquisition Fund. Funding for locating nests in 1997 was provided by the Illinois Council on Food and Agricultural Research (C-FAR).

10 August 1998; accepted 9 October 1998

Guiding Questions for Reading This Article

A. About the Article

1. Name the journal and the year in which this study was published.

2. The authors of the paper are from several agencies and a university. Name three of the institutions they represent.

3. What is the genus, species, and subspecies name of the greater prairie chicken? What is its habitat?

4. Specialized vocabulary: Write a brief definition of each term.

   Described in the article text or end notes:

   extinction vortex

   microsatellite markers

   Not defined in the article:

   allele

   clutch

   extirpation

   lek

   radiotelemetry

   recruitment

B. About the Study

5. State why loss of genetic variation may harm a population and why chance events harm small populations more than large populations.

6. What, according to the authors, is the single most important factor leading to the decline in populations of greater prairie chickens?

7. From the information at the top of p. 1696, summarize the changes in counts of male greater prairie chickens from 1963 to 1994 in the study area in Jasper County, Illinois.

8. What is the *y*-axis label on the left side of Figure 2, and what symbols on the graph refer to annual means for those data? What are the highest and lowest values for these data in the range of years between 1963 and 1990? To what do the vertical bars and the numbers at the bottom of those bars refer?

9. In determining clutch size, fertility, and egg success, what were some of the factors (disturbance, etc.) that the investigators had to exclude or consider in their analysis? See note 24 about the field methods on pp. 1697–1698.

10. How did managers manipulate the population, beginning in August 1992? What led them to do this manipulation, and what were their objectives?

11. What were the results in the first few years after the transplantation?

12. In Figure 2, compare (a) the number of males (triangles, right *y*-axis) and (b) the percentage of eggs hatched (circles, left *y*-axis) for 1987–1989 and 1995–1997.

13. In Figure 2, which parameter—(a) number of males (triangles) or (b) percentage of eggs hatched (circles)—is a better indicator of the population size? Why? Which of those two parameters is a better indicator of the genetic diversity? Why?

14. What are some climatic factors the authors considered that could have affected the breeding success of greater prairie chickens?

15. The authors state they have no information that might have served as controls for data from the 1993, 1994, and 1997 seasons when breeding individuals included several introduced from other states. What clutches would have served as controls for this comparison?

16. Is this a *field study,* with data collected on organisms in their natural habitat, or is this a *lab study,* in which organisms are studied under controlled conditions in the laboratory?

17. Is this an *observational study,* in which quantitative, observational data are taken but no experimental manipulation is made, or is this an *experimental study,* in which researchers make manipulations by which the effects of different variables are tested, one at a time?

C. General Conclusions and Extensions of the Work

18. Both fitness and genetic diversity declined in the Illinois prairie chickens over a 32-year period. After seeing breeding results following bird translocations, the authors conclude which of the following is more critical to maintaining populations of prairie chickens—sufficient population size or sufficient genetic diversity? Explain the basis of their conclusion.

19. The two variables contributing to the extinction vortex—decreased population size and decreased genetic diversity—are difficult to separate in this study system. Can you think of a way to investigate their separate effects on reproduction in prairie chicken populations?

20. Imagine that you were a member of this research team. What could be a possible follow-up test that extends this work? Briefly state a different experiment or measurement you would do within this research system.

Note: For further reading about conservation efforts with other endangered North American bird species, you might look for information about the following: California condor, Eskimo curlew, Kirtland's warbler, trumpeter swan, red-cockaded woodpecker, whooping crane.

ARTICLE 8

Inquiry Figure 9*: *Is the Rotation of the Internal Rod in ATP Synthase Responsible for ATP Synthesis?*

Introduction—The Article and Phenomenon Under Study

ATP is made from ADP and inorganic phosphate, using the energy from proton flow through the membrane protein complex known as ATP synthase. A particular subunit of the protein complex acts as a rotary motor in the production of ATP. Inquiry Figure 9 Campbell/Reece *Biology,* Eighth Edition, shows how investigators provided direct evidence for this mechanism of ATP synthesis by means of rotation in the following article:

H. Itoh et al., Mechanically driven ATP synthesis by F$_1$-ATPase, *Nature* 427:465–468 (2004).

Read the complete article beginning on the next page and then answer the questions following the article.

---

* This Inquiry Figure appeared in Campbell/Reece *Biology,* Eighth Edition, as Inquiry Figure 9.15.

▼ **Figure 9**

INQUIRY

Is the rotation of the internal rod in ATP synthase responsible for ATP synthesis?

EXPERIMENT Previous experiments on ATP synthase had demonstrated that the "internal rod" rotated when ATP was hydrolyzed. Hiroyasu Itoh and colleagues set out to investigate whether simply rotating the rod in the opposite direction would cause ATP synthesis to occur. They isolated the internal rod and catalytic knob, which was then anchored to a nickel plate. A magnetic bead was bound to the rod. This complex was placed in a chamber containing an array of electromagnets, and the bead was manipulated by the sequential activation of the magnets to rotate the internal rod in either direction. The investigators hypothesized that if the bead were rotated in the direction opposite to that observed during hydrolysis, ATP synthesis would occur. ATP levels were monitored by a "reporter enzyme" in the solution that emits a discrete amount of light (a photon) when it cleaves ATP. Their hypothesis was that rotation in one direction would result in more photons than rotation in the other direction or no rotation at all.

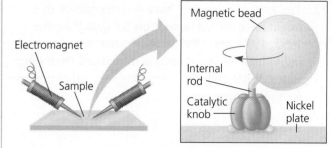

RESULTS More photons were emitted by spinning the rod for 5 minutes in one direction (yellow bars) than by no rotation (gray bars) or rotation in the opposite direction (blue bars).

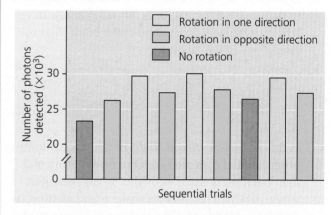

CONCLUSION The researchers concluded that the mechanical rotation of the internal rod in a particular direction within ATP synthase appears to be all that is required for generating ATP. As ATP synthase is the smallest rotary motor known, one of the goals in this type of research is to learn how to use its activity in artificial ways.

SOURCE H. Itoh et al., Mechanically driven ATP synthesis by F$_1$-ATPase, *Nature* 427:465–468 (2004).

INQUIRY IN ACTION Read and analyze the original paper in *Inquiry in Action: Interpreting Scientific Papers.*

WHAT IF? The "no rotation" (gray) bars represent the background level of ATP in the experiment. When the enzyme is rotated one way (yellow bars), the increase in ATP level suggests synthesis is occurring. For enzymes rotating the other way (blue bars), what level of ATP would you expect compared to the gray bars? (Note: this may not be what is observed.)

letters to nature

••

Mechanically driven ATP synthesis by F₁-ATPase

Hiroyasu Itoh[1,2], Akira Takahashi[3], Kengo Adachi[4], Hiroyuki Noji[5], Ryohei Yasuda[6], Masasuke Yoshida[7] & Kazuhiko Kinosita Jr[4]

[1]*Tsukuba Research Laboratory, Hamamatsu Photonics KK, and* [2]*CREST "Creation and application of soft nano-machine, the hyperfunctional molecular machine" Team 13*, Tokodai, Tsukuba 300-2635, Japan*
[3]*System Division, Hamamatsu Photonics KK, Joko, Hamamatsu 431-3103, Japan*
[4]*Center for Integrative Bioscience, Okazaki National Research Institutes, Okazaki 444-8585, Japan*
[5]*Institute of Industrial Science, University of Tokyo, Tokyo 153-8505, Japan*
[6]*Cold Spring Harbor Laboratory, Cold Spring Harbor, New York 11724 USA*
[7]*ERATO "ATP System", 5800-3 Nagatsuta, Yokohama 226-0026, Japan*

ATP, the main biological energy currency, is synthesized from ADP and inorganic phosphate by ATP synthase in an energy-requiring reaction[1-3]. The F₁ portion of ATP synthase, also known as F₁-ATPase, functions as a rotary molecular motor: *in vitro* its γ-subunit rotates[4] against the surrounding $\alpha_3\beta_3$ subunits[5], hydrolysing ATP in three separate catalytic sites on the β-subunits. It is widely believed that reverse rotation of the γ-subunit, driven by proton flow through the associated F₀ portion of ATP synthase, leads to ATP synthesis in biological systems[1-3,6,7]. Here we present direct evidence for the chemical synthesis of ATP driven by mechanical energy. We attached a magnetic bead to the γ-subunit of isolated F₁ on a glass surface, and rotated the bead using electrical magnets. Rotation in the appropriate direction resulted in the appearance of ATP in the medium as detected by the luciferase–luciferin reaction. This shows that a vectorial force (torque) working at one particular point on a protein machine can influence a chemical reaction occurring in physically remote catalytic sites, driving the reaction far from equilibrium.

When isolated F₁ hydrolyses ATP, its central γ-subunit rotates anticlockwise[4] when viewed from above in Fig. 1a, with an efficiency of chemical-to-mechanical energy conversion approaching 100% (ref. 8). The purpose of this study was to show that the chemomechanical coupling in the F₁ motor is completely reversible, and that reversal is achieved by manipulating a single variable—that is, the rotary angle of the γ-subunit. Any molecular machine would be reversible if one could manipulate all constituent atoms at will. Whether one or a few thermodynamic handles exist in a chemomechanical molecular machine such that its operation can be controlled through that handle in both directions is an important but unresolved issue. For example, whether one can synthesize ATP by pulling back a linear molecular motor such as myosin or kinesin—and if so, where to pull—is unknown. Reversal of the whole ATP synthase is well documented[1,9], including the demonstration of γ-subunit reorientation under synthesis conditions[10], but whether or not the γ-subunit angle serves as a single handle for F₁ reversal has not been tested.

To prove this reversibility, we used $\alpha_3\beta_3\gamma$, the minimal subcomplex of F₁ that shows ATP-catalysed rotation[4]. The subcomplex was attached to a glass surface through histidine residues engineered at the amino terminus of the β-subunits, and a magnetic bead coated with streptavidin was attached to the γ-subunit, which had been biotinylated at two engineered cysteines (Fig. 1a). The beads were rotated with magnets (Fig. 1b–d) in a medium containing ADP and phosphate as substrates and the luciferin–luciferase system[11,12], which emits a photon when it captures and hydrolyses ATP. The initial idea was to count these chemiluminescent photons (Fig. 1b); however, background luminescence originating from contaminant ATP present in ADP even after purification was a problem. Thus, the

letters to nature

volume of medium per active F_1 molecule had to be small.

First, we tried to reduce the volume by making microdroplets in oil (Fig. 2a). Figure 2b shows data from a 4×4 array of droplets in one chamber. The beads in droplets were rotated at 10 Hz alternately for 5 min each in the direction of hydrolysis (anticlockwise when viewed from top in Fig. 1a) and synthesis (clockwise). As seen in Fig. 2b, 14 out of 16 droplet curves showed the M-shaped pattern of photon counts expected for the sequence of rotation direction when the overall decline was taken into account. The decline was due to the gradual disappearance of the aqueous phase into oil: although we saturated the oil with water before the experiment, droplets tended to shrink over time. For random photon counts, the probability of observing a slanting M shape is 8^{-1}. The probability of observing 14 or more M shapes out of 16 is 2×10^{-11}. The data set shown in Fig. 2b thus strongly indicates mechanical synthesis. The experiment is extremely difficult (at most a few beads rotate in each droplet), however, and we have obtained only a few more data sets that contained several M-shaped patterns.

We thus tried to increase the number of rotating beads in an ordinary observation chamber (Fig. 1c) by infusing a concentrated solution of beads carrying F_1. To allow F_1 to rotate in the proper direction, we derivatized only the bottom surface of the chamber with nickel nitrilo-triacetic acid (Ni^{2+}-NTA), which would specifically bind the β-histidines. In control experiments done in 4 mM ATP (no ADP) and without magnets, we found in the field of view of $1.0 \times 10^5\, \mu m^2$ as many as $480 \pm 70\ F_1$ molecules rotating anticlockwise at the bottom (three chambers). However, the high density of beads resulted in nonspecific binding to the ceiling, where 100 ± 20 beads rotated clockwise (as viewed from above the chamber). We also tested in the ATP medium whether forced rotation by external magnets would damage F_1. After confirming ATP-driven rotation, we turned on the magnets and applied several bursts of hundreds of revolutions at 10 Hz in both directions. When the magnets were turned off, ATP-driven rotation resumed.

Synthesis was shown in the ADP–luciferase medium by accumulating chemiluminescence photons over a series of 5-min intervals in which the magnetic field was rotated at 10 Hz in either direction or turned off. All series produced the expected pattern (Fig. 3a): higher photon counts during clockwise rotation (S, synthesis) than during anticlockwise rotation (H, hydrolysis) or no rotation (N). This graph compiles all data taken in consecutive experiments (about half of the experiments failed at some point, for example during chamber preparation or because of a large focus drift, and did not produce data). Mechanical synthesis in the flat chamber was reproducible, although variation among data was still large.

We note that in most curves shown in Fig. 3a including the total counts, counts during anticlockwise rotation (H) are higher than those at no rotation (N). This is due to the presence of F_1 at the ceiling of the chamber as stated above. For these upside-down F_1 molecules, anticlockwise bead rotation will drive ATP synthesis. Indeed, when we flipped the chamber upside down after obtaining the unbroken blue line in Fig. 3a, the count pattern was reversed, as shown by the broken blue line. Another reason for the higher counts during anticlockwise rotation was that luciferase did not consume all of the newly synthesized ATP in 5 min: as shown by the unbroken lines in Fig. 3b, the luminescence at no rotation was high after ATP synthesis at the arrows. Luminescence decay after ATP mixing was shown to involve a component with a lifetime of about 3 min (see Supplementary Information). Taking all of the above points into account, we consider that mechanical synthesis has been conclusively demonstrated.

Figure 4 shows the effect of rotary speed on the efficiency of ATP synthesis. The synthesis rate apparently saturated above 3 Hz (Fig. 4b). This was because larger beads or bead aggregates failed to rotate at high speeds, as confirmed by direct observation (uncoupling between magnet and bead rotations). Calibration of the

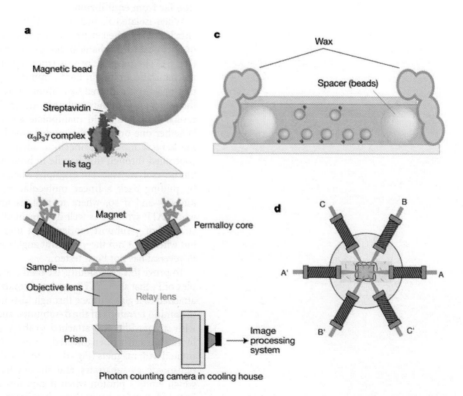

Figure 1 Experimental set-up. **a**, Basic design. The structures of F_1 and streptavidin are from ref. 21 and ref. 22, respectively. The bead is not to scale and the orientation of streptavidin is unknown. **b**, Side view of the optical system. **c**, Observation chamber. Ni^{2+}-NTA was applied only to the bottom surface, but some F_1-conjugated beads rotated on the ceiling. **d**, Top view of the magnets.

letters to nature

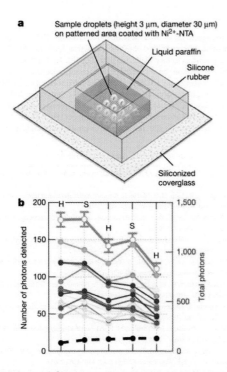

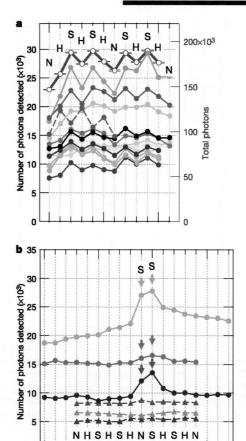

Figure 2 Rotational synthesis in microdroplets. **a**, Observation chamber. On the silanized bottom coverglass, 4×4 spots separated by $\sim50\,\mu m$ and measuring $20{-}30\,\mu m$ were derivatized with Ni^{2+}-NTA, which rendered the spots hydrophilic. An oil layer with a thickness of $\sim3\,mm$ was placed on top. Using a glass micropipette, we formed a droplet of $\sim1\,pl$ containing F_1 on each spot and then replaced the solution with one containing beads, ADP, phosphate and the chemiluminescence system. **b**, Simultaneous observation of 16 droplets. Thick magenta curve with open circles represents the sum of all unbroken curves; error bars represent $\pm\,2\sigma$, where σ is the expected s.d. for photon statistics (square root of the total count). The broken curve at the bottom represents a control in which an area outside the droplets was imaged. See Supplementary Information for details.

Figure 3 Rotational synthesis in a flat chamber. Each symbol shows the number of photons detected over 5 min in an image area of $7.8 \times 10^4\,\mu m^2$. N, no rotation; H, 10-Hz rotation in the hydrolysis direction (for F_1 on the bottom of the chamber); S, 10-Hz rotation in the synthesis direction. **a**, Results from 12 chambers, distinguished by different colours. Broken blue curve represents results obtained after the chamber giving rise to the unbroken blue curve was flipped upside down. Thick magenta curve with open circles represents the sum of all unbroken curves; error bars represent $\pm\,4\sigma$ (>99.99% confidence). For clarity, some curves have been shifted vertically within $\pm\,1,000$ counts. **b**, Control experiments. Unbroken curves show results either without rotation or with imposed clockwise rotation for synthesis at 10 Hz (arrows). Broken curves show experiments carried out as in **a** except that phosphate was omitted from the medium. The counts in the broken curves are low in comparison to the others because phosphate tended to increase the background, apparently by increasing the portion of luciferase that reacted with ATP extremely slowly (see Supplementary Information).

photon-to-ATP ratio (see Supplementary Information) indicated a synthesis rate of about five molecules of ATP per second under rotation at 3 Hz, as compared with the nine molecules of ATP per second expected for the one ATP molecule per 120° scheme[8].

Although Fig. 4 represents our best data so far and variation among chambers is large (Fig. 3a), we anticipate that the coupling between mechanical rotation of the γ-subunit and chemical synthesis is tight, at least at low speeds. The significant synthesis at low speeds (Fig. 4), which were much lower than the maximal rotary speed of 130 Hz of this motor during ATP hydrolysis[13], merits attention. Because the slow rotation was at a constant speed, it is likely that chemical reactions were at quasi-equilibrium at all angles. Rotation by ATP hydrolysis can also be made slow and at a constant speed by attaching a long actin filament to the γ-subunit[8], again suggesting quasi-equilibrium. The implication is that ATP synthesis in F_1 proceeds as a straightforward reversal of the hydrolysis reaction, tracking the same reaction pathway in the opposite direction. On that pathway, both hydrolysis and synthesis reactions are controlled by one mechanical handle, the rotary angle of the γ-subunit. The situation contrasts with the more complex rotary motor of bacterial flagella, where rotational directions can be switched without reversing the proton motive force[14].

ATP synthesis is a chemical reaction that is energetically uphill, requiring 80–100 pN nm of energy under physiological conditions[2]. In the experiments shown here, contaminant ATP amounted to about 1 nM, implying that roughly 30 pN nm of free energy was needed per molecule of ATP synthesized (see Supplementary

Information). If F_1, or the motor enzyme myosin, is mixed with high concentrations of ADP and phosphate in the absence of ATP, some ATP is spontaneously formed on the enzyme without input of energy[15–18]. This, however, is a dead-end reaction and the ATP that has been formed is not available in the medium: release of the tightly bound ATP requires an external supply of energy[1,13].

Here we have demonstrated repetitive synthesis by F_1 ($\sim10^3$ ATP molecules in 5 min), leading to appearance of the product in the medium. To our knowledge, this is the first accomplishment of artificial chemical synthesis by a vectorial force (although nature presumably has been doing this for millions of years). Pressure could also shift a chemical equilibrium by acting on substrates (and solvent); however, in our experiments the chemical equilibrium *per se* is on the side of almost complete hydrolysis, and a force on a point remote from substrates counters the hydrolysis reaction and pushes the equilibrium to the point of favouring synthesis. Because we still have to rely on nature's nanomachine, the F_1 motor, our

letters to nature

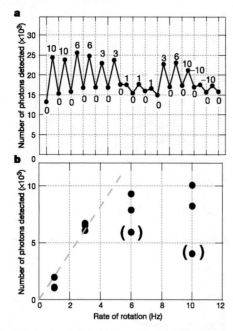

Figure 4 Dependence of synthesis efficiency on the rate of magnet rotation. **a**, Photons detected over an image area of $1.0 \times 10^5 \, \mu m^2$ in 5-min periods at various rotary speeds (shown in Hz; negative numbers indicate rotation in the hydrolysis direction). **b**, Photon increments during rotation. From the photon count for each synthesis rotation in **a**, the average of the adjacent no-rotation counts was subtracted. Symbols in the parentheses show the last two measurements for synthesis in **a**; low values for these presumably reflect sample deterioration.

primary goal is to understand fully how it works and thereby to exploit the mechanism in artificial ways. Improving the present system for more quantitative assays, such as the torque and speed dependence of the coupling efficiency, that can be compared with theoretical predictions[7,19] should be our next task. The key is to obtain magnetic beads that are small and uniform in size. □

Methods

Materials

A mutant $\alpha_3\beta_3\gamma$ subcomplex (comprising C193S α-, His_{10}-β-, and S107C, I210C γ-subunits; referred to as F_1 in this paper), derived from the thermophilic *Bacillus* strain PS3, was biotinylated at the only cysteines on the γ-subunit (ref. 8). Streptavidin-coated magnetic beads (Seradyn; nominally 0.7 μm) were lightly centrifuged to remove large beads and aggregates (but elimination was incomplete). Biotinylated F_1 (400 pM) was incubated with beads (~50 pM) in buffer A (50 mM 3-(N-morpholino)propanesulphonic acid/KOH, 20 mM K_2SO_4, 4 mM $MgSO_4$, pH 7.6) for 10 min at 23 °C, washed with buffer A, and concentrated with a magnet. Luciferase[12] was a kind gift from Kikkoman Co. We purified ADP (K-salt; Sigma) as described[20] on a Poros HQ/L column (Applied Biosystems).

Observation chamber

For the chamber shown in Fig. 1c, a $32 \times 24 \, mm^2$ coverglass was functionalized by a silane coupling agent with an SH group (TSL 8380; GE Toshiba Silicone), and reacted first with 10 mg ml^{-1} maleimide-C_3-NTA (Dojindo) and then with 10 mM $NiCl_2$. Two parallel strips of Lumirror polyester film (TORAY) were placed on the coverglass, and a $18 \times 18 \, mm^2$ coverglass coated with hexamethyldisilazane was placed on top to form a flow chamber. We infused ~500 pM F_1-conjugated magnetic beads in buffer A_3 (buffer A containing 3 mg ml^{-1} bovine serum albumin (BSA)) and incubated the chamber for 30 min at 23 °C. After infusing buffer A_3 several times, buffer A containing 10 μM luciferase, 1 mM luciferin, 10 mM K_2PO_4, 200 μM ADP and 1.5 mg ml^{-1} BSA, together with a small number of 3-μm Dynabeads M-280 beads (Dynal) that would eventually serve as spacers, was infused. The spacer strips were carefully removed and the upper coverglass was pressed to reduce the chamber height to ~3 μm, as determined by the spacer beads. The chamber was then sealed with wax and subjected to observation.

Microscopy

The chamber was placed on an inverted ICM 450 microscope (Zeiss). Luminescence was collected with an oil immersion ×60 objective, numerical aperture 1.45 (Olympus) and deflected with a prism to a factory-made side port (Fig. 1b). The beam was focused with an ED Plan ×2 objective (Nikon) onto a cooled photon-counting camera (V8070U-64-N230/C4566 equipped with an Argus 50 image processing system; Hamamatsu Photonics), which recorded centroid positions of incoming photons (counts in the dark: ~15 s^{-1} over the whole image plane). To rotate the beads, we placed three opposing pairs of custom-made electromagnets with a Permalloy core on the chamber (Fig. 1d). The three pairs were activated with a custom circuit 120° out of phase to produce a rotating magnetic field (in either direction).

Received 5 May; accepted 31 October 2003; doi:10.1038/nature02212.

1. Boyer, P. D. The ATP synthase—a splendid molecular machine. *Annu. Rev. Biochem.* **66**, 717–749 (1997).
2. Kinosita, K. Jr, Yasuda, R., Noji, H. & Adachi, K. A rotary molecular motor that can work at near 100% efficiency. *Phil. Trans. R. Soc. Lond. B* **355**, 473–489 (2000).
3. Yoshida, M., Muneyuki, E. & Hisabori, T. ATP synthase—a marvellous rotary engine of the cell. *Nature Rev. Mol. Cell Biol.* **2**, 669–677 (2001).
4. Noji, H., Yasuda, R., Yoshida, M. & Kinosita, K. Jr Direct observation of the rotation of F_1-ATPase. *Nature* **386**, 299–302 (1997).
5. Abrahams, J. P., Leslie, A. G. W., Lutter, R. & Walker, J. E. Structure at 2.8 Å resolution of F_1-ATPase from bovine heart mitochondria. *Nature* **370**, 621–628 (1994).
6. Boyer, P. D. & Kohlbrenner, W. E. in *Energy Coupling in Photosynthesis* (eds Selman, B. R. & Selman-Reimer, S.) 231–240 (Elsevier, Amsterdam, 1981).
7. Oosawa, F. & Hayashi, S. The loose coupling mechanism in molecular machines of living cells. *Adv. Biophys.* **22**, 151–183 (1986).
8. Yasuda, R., Noji, H., Kinosita, K. Jr & Yoshida, M. F_1-ATPase is a highly efficient molecular motor that rotates with discrete 120° steps. *Cell* **93**, 1117–1124 (1998).
9. Turina, P., Samoray, D. & Gräber, P. H$^+$/ATP ratio of proton transport-coupled ATP synthesis and hydrolysis catalysed by CF_0F_1-liposomes. *EMBO J.* **22**, 418–426 (2003).
10. Zhou, Y., Duncan, T. M. & Cross, R. L. Subunit rotation in *Escherichia coli* F_0F_1–ATP synthase during oxidative phosphorylation. *Proc. Natl Acad. Sci. USA* **94**, 10583–10587 (1997).
11. McElroy, W. D., Seliger, H. H. & White, E. H. Mechanism of bioluminescence, chemiluminescence and enzyme function in the oxidation of firefly luciferin. *Photochem. Photobiol.* **10**, 153–170 (1969).
12. Hattori, N., Kajiyama, N., Maeda, M. & Murakami, S. Mutant luciferase enzymes from fireflies with increased resistance to benzalkonium chloride. *Biosci. Biotechnol. Biochem.* **66**, 2587–2593 (2002).
13. Yasuda, R., Noji, H., Yoshida, M., Kinosita, K. Jr & Itoh, H. Resolution of distinct rotational substeps by submillisecond kinetic analysis of F_1-ATPase. *Nature* **410**, 898–904 (2001).
14. Berg, H. C. The rotary motor of bacterial flagella. *Annu. Rev. Biochem.* **72**, 19–54 (2003).
15. Yoshida, M. The synthesis of enzyme-bound ATP by the F_1-ATPase from the thermophilic bacterium PS3 in 50% dimethylsulfoxide. *Biochem. Biophys. Res. Commun.* **114**, 907–912 (1983).
16. Sakamoto, J. Effect of dimethylsulfoxide on ATP synthesis by mitochondrial soluble F_1-ATPase. *J. Biochem.* **96**, 483–487 (1984).
17. Wolcott, R. G. & Boyer, P. D. The reversal of the myosin and actomyosin ATPase reactions and the free energy of ATP binding to myosin. *Biochem. Biophys. Res. Commun.* **57**, 709–716 (1974).
18. Mannherz, H. G., Schenck, H. & Goody, R. S. Synthesis of ATP from ADP and inorganic phosphate at the myosin-subfragment 1 active site. *Eur. J. Biochem.* **48**, 287–295 (1974).
19. Wang, H. & Oster, G. Energy transduction in the F_1 motor of ATP synthase. *Nature* **396**, 279–282 (1998).
20. Oishi, N. & Sugi, H. *In vitro* ATP-dependent F-actin sliding on myosin is not infuenced by substitution or removal of bound nucleotide. *Biochim. Biophys. Acta* **1185**, 346–349 (1994).
21. Menz, R. I., Walker, J. E. & Leslie, A. G. W. Structure of bovine mitochondrial F_1-ATPase with nucleotide bound to all three catalytic sites: implications for the mechanism of rotary catalysis. *Cell* **106**, 331–341 (2001).
22. Freitag, S., Trong, I. L., Klumb, L., Stayton, P. S. & Stenkamp, R. E. Structural studies of the streptavidin binding loop. *Protein Sci.* **6**, 1157–1166 (1997).

Supplementary Information accompanies the paper on **www.nature.com/nature**.

Acknowledgements We thank T. Hayakawa and T. Hiruma of Hamamatsu Photonics KK who allowed H.I. to work on this project for more than 6 years; S. Brenner for the idea of using microdroplets; M. Sugai for initial work; M. Shio, members of the former CREST Team 13 and the current Kinosita and Yoshida laboratories for help and advice; I. Mizuno, K. Suzuki, S. Uchiyama and Y. Mizuguchi for the photon-counting system; C. Gosse and H. Miyajima for the magnetic tweezers; K. Abe and K. Rikukawa for microscopy; S. Murakami for luciferase; and H. Umezawa and M. Fukatsu for laboratory management. This work was supported in part by Grants-in-Aid from the Ministry of Education, Culture, Sports, Science and Technology of Japan, and Burroughs Wellcome Fund (R.Y.).

Competing interests statement The authors declare that they have no competing financial interests.

Correspondence and requests for materials should be addressed to H.I. (hiritoh@hpk.trc-net.co.jp).

Guiding Questions for Reading This Article

A. About the Article

1. What is the name of the first author of this research team?

2. Most of the authors of this paper work in major research laboratories in what nation? What is the name of the author affiliated with the Cold Spring Harbor Laboratory in the United States?

3. What is the name of the journal in which this article was published? This paper was published in what month and year?

4. This research was supported in part by grants from what organization? (See Acknowledgements, p. 468.)

5. The article's abstract (p. 465, first paragraph, in bold print) summarizes the major contributions of the study. Write the one sentence from the abstract that states the basic concept for which the study provides direct evidence.

6. Specialized vocabulary: Write a brief definition of each term.

   chemiluminescence

   hydrolysis

   luciferase

   photon

B. About the Study

7. Was this an *in vivo study* (in intact cells, in living organisms) or an *in vitro study* (literally, "in glass," meaning in test tubes or in laboratory containers)?

8. The action of isolated F1 particles of ATP synthase appeared to be reversible. (a) The central subunit rotates *anticlockwise (that is, counterclockwise)* when what happens—A or B? (A) ATP is synthesized; (B) ATP is hydrolyzed (broken down to ADP and inorganic P). (b) The central subunit rotates *clockwise* when what happens—A or B? (A) ATP is synthesized; (B) ATP is hydrolyzed (broken down to ADP and inorganic P).

9. The researchers manipulated a single subunit of the ATP synthase complex, the γ (gamma) subunit. Prior to this study, what aspect of the γ subunit in F1 reversal had not yet been tested?

10. One end of the F1 test complex was attached to a glass slide by modifying two cysteine amino acids of one end of the complex. The other end of the F1 test complex, the γ end, was attached to a magnetic bead using a specific attachment protein (streptavidin). Examine the experimental setup in Figure 1 (p. 466). Although the drawing in Figure 1a is not to scale, which is the larger element, the F1 protein complex or the magnetic bead? How did the investigators manipulate this system to cause the γ end of the complex to rotate?

11. The investigators noted that during ATP hydrolysis, the complex rotated in one direction. What reaction did they hypothesize would occur if the complex was rotated in the opposite direction?

12. The investigators used photons of light produced in a chemiluminescent luciferin-luciferase system as an indicator of ATP synthesis. When ATP is present, this enzyme system will hydrolyze ATP to produce ADP and inorganic phosphate and release photons of light. (Thus, it serves as a "reporter enzyme," as mentioned in Inquiry Figure 9.) What does the y-axis on the left side of Figures 3a and 3b represent? This is an indirect measure of the rate of which rotation-driven process—ATP synthesis or ATP hydrolysis?

13. Figure 3a shows the number of photons detected with respect to rotation of the γ-subunit of the ATP synthase. N, S, and H refer to what three rotation states? In Figure 3a, the relative number of photons detected is highest at which type of rotation—N, S, or H?

14. In the control experiments graphed in Figure 3b, the broken lines show the lowest number of photons detected. What was different about those experiments to produce such low numbers? Why did this omission serve as a control for the experiment?

C. General Conclusions and Extensions of the Work

15. What new information about the role of the γ subunit in ATP synthase did this research project elucidate? How does the subunit serve as a reversible motor?

16. What future tasks do the authors wish to do once they understand this motor mechanism fully?

Inquiry Figure 30*: *Can Flower Shape Influence Speciation Rate?*

Introduction—The Article and Phenomenon Under Study

What evolutionary processes contribute to species diversity? In order for a new species to evolve, there must be genetic divergence and reproductive isolation. Based on current eco-logical observations and on correlations in the fossil record, we assume the great diversity of flowering plant species is related to animal diversity and to plant-pollinator relation-ships. The investigator in this paper took a unique approach in testing the hypothesis that flower shape can influence the rate at which new species form. The study compared sister groups of angiosperm families to see the relationship between traits that promote reproductive isolation and increased species diversification. This study, from the following paper, is the subject of Inquiry Figure 30 Campbell/Reece *Biology*, Eighth Edition:

R. D. Sargent, Floral symmetry affects speciation rates in angiosperms, *Proceedings of the Royal Society B: Biological Sciences,* 271:603–608 (2004).

Read the complete article beginning on the next page and then answer the questions following the article.

---

* This Inquiry Figure appeared in Cambell/Reece *Biology*, Eighth Edition, as Inquiry Figure 30.14.

▼ **Figure 30** **INQUIRY**

Can flower shape influence speciation rate?

EXPERIMENT The shapes of bilaterally symmetrical flowers can force pollinators seeking nectar to enter the flower from certain directions only. This facilitates the placement of pollen on particular parts of the pollina-tor's body, which in turn can increase how effectively pollen is transferred between flowers of the same species. Such specificity in plant-pollinator relationships tends to decrease gene flow between diverging plant popu-lations and hence could lead to increased rates of plant speciation. Risa Sargent, then at the University of British Columbia, tested this hypothesis by determining whether angiosperm families with bilaterally symmetrical flowers contain more species than do closely related families with radially symmetrical flowers. She took this approach:

❶ Identify as many cases as possible in which a clade of one or more families with bilaterally symmetrical flowers is sister to a clade whose members have radially symmetrical flowers. By definition, such sister clades share an immediate common ancestor and hence had the same amount of time to form new species.

❷ Determine the number of species in each pair of sister clades. The hypothesis would be supported if bilaterally symmetrical clades consistently have more species than their radially symmetrical sister clades.

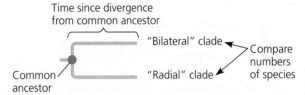

RESULTS Sargent identified 19 pairs of sister clades that differed in flower shape. In 15 of these pairs, clades with bilateral symmetry had more species than did clades with radial symmetry.

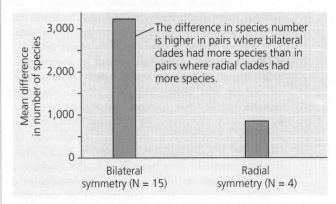

CONCLUSION Sargent's results supported the hypothesis that flower shape affects the rate at which new species form, with bilateral symme-try promoting increased rates of speciation.

SOURCE R. D. Sargent, Floral symmetry affects speciation rates in an-giosperms, *Proceedings of the Royal Society B: Biological Sciences* 271:603–608 (2004).

INQUIRY IN ACTION Read and analyze the original paper in *Inquiry in Action: Interpreting Scientific Papers.*

WHAT IF? Does this study show that flower shape is *correlated with* the rate at which new species form, or that flower shape is *responsible for* this rate? Explain.

THE ROYAL
SOCIETY

Received 25 July 2003
Accepted 13 November 2003
Published online 16 February 2004

Floral symmetry affects speciation rates in angiosperms

Risa D. Sargent

Department of Zoology, University of British Columbia, 6270 University Boulevard, Vancouver, British Columbia V6T 1Z4, Canada (sargent@zoology.ubc.ca)

Despite much recent activity in the field of pollination biology, the extent to which animal pollinators drive the formation of new angiosperm species remains unresolved. One problem has been identifying floral adaptations that promote reproductive isolation. The evolution of a bilaterally symmetrical corolla restricts the direction of approach and movement of pollinators on and between flowers. Restricting pollinators to approaching a flower from a single direction facilitates specific placement of pollen on the pollinator. When coupled with pollinator constancy, precise pollen placement can increase the probability that pollen grains reach a compatible stigma. This has the potential to generate reproductive isolation between species, because mutations that cause changes in the placement of pollen on the pollinator may decrease gene flow between incipient species. I predict that animal-pollinated lineages that possess bilaterally symmetrical flowers should have higher speciation rates than lineages possessing radially symmetrical flowers. Using sister-group comparisons I demonstrate that bilaterally symmetric lineages tend to be more species rich than their radially symmetrical sister lineages. This study supports an important role for pollinator-mediated speciation and demonstrates that floral morphology plays a key role in angiosperm speciation.

Keywords: reproductive isolation; pollination; sister group comparison; zygomorphy

1. INTRODUCTION

One of the fundamental objectives of evolutionary biology is to understand why there are such vast differences in speciation rates across taxonomic lineages (Futuyma 1998). The biological species concept emphasizes reproductive isolation as the key factor in speciation. Consequently, traits that promote reproductive isolation among adjacent populations are considered key to the origin of new species (Grant & Grant 1965; Schluter 2001).

One prominent evolutionary trend in flowering plants is the fusion of petals and overall reduction in the number of stamens and carpels (Endress 1997). The adaptive explanation for these changes is that they have allowed more precise pollination by specialist insect pollinators and, consequently, less expense of pollen and nectar (Regal 1977; Takhtajan 1991). From the plant's perspective, the selective advantage of specialist pollination is clear; plants are less likely to receive incompatible pollen or to have their pollen transferred to an incompatible stigma. Indeed, selection for pollinator specialization has been invoked to explain divergence in several floral traits including: animal pollination, nectar guides, nectar spurs, bilateral symmetry and secondary pollen presentation (Bawa 1995; Waser 2001). Grant (1949) suggested that in the angiosperms, floral morphology has diverged more rapidly than vegetative characteristics, explaining its widespread preference as a basis for taxonomic classification. Many authors suggest that this divergence has been driven largely by selection via pollinators (Grant 1949, 1994; Stebbins 1970; Faegri & van der Pijl 1979; however, see Waser 1998, 2001). Accordingly, the occurrence of animal pollination has been invoked to explain differences in diversification rates across angiosperm lineages (Eriksson & Bremer 1992; Dodd et al. 1999).

The importance of pollinator-mediated selection in angiosperms is well supported by theory (Kiester et al. 1984) and experimental data (Galen 1996). In the genus *Mimulus*, evidence suggests that discrimination by pollinators (bees and hummingbirds) is responsible for reproductive isolation between two sympatric species (Schemske & Bradshaw 1999). In the genus *Aquilegia*, differences in the form of nectar spurs are correlated with differences in pollinators that visit a flower; the size and placement of the spurs affect reproductive isolation by reducing visitation by some pollinators and increasing visitation by others (Hodges & Arnold 1994). The presence of spurs has also been shown to correlate with the degree of diversification in other clades, supporting the hypothesis that they play a general role in reproductive isolation (Hodges & Arnold 1995).

Floral symmetry was one of the earliest traits used to relate morphology to function in the pollination of angiosperms (Neal et al. 1998). There are two main forms of symmetry described in the angiosperms: bilateral symmetry (zygomorphy) and radial symmetry (actinomorphy). Actinomorphy is considered to be the ancestral form (Takhtajan 1969) with zygomorphy having originated several times independently (Takhtajan 1991; Donoghue et al. 1998). Several theories have been proposed for the adaptive significance of zygomorphy (reviewed by Neal et al. 1998). The pollen position hypothesis posits that in zygomorphic flowers, pollinators are restricted in the directionality of approach and movement within and between flowers (Leppik 1972; Ostler & Harper 1978; Cronk & Moller 1997). By contrast, actinomorphic flowers can be approached from any direction and are not able to restrict pollinator movement within the flower. Hence, in zygomorphic flowers the specificity of pollen placement is greatly improved. Once precise placement of pollen on the pollinator is achieved, reproductive isolation is possible.

Proc. R. Soc. Lond. B (2004) **271**, 603–608
DOI 10.1098/rspb.2003.2644

603

Wherever a trait change has occurred convergently in several lineages there is an opportunity to compare the resulting differences in diversity between the lineage and its sister lineage (reviewed by Barraclough *et al.* 1998). Given sufficient comparisons one can test the hypothesis that the evolution of the trait has had a consistent, replicable effect on diversification. Several studies have examined the impact of different traits on diversification rates in angiosperms (e.g. Farrell *et al.* 1991; Hodges & Arnold 1995; Dodd *et al.* 1999; Heilbuth 2000; Barraclough & Savolainen 2001; Verdu 2002) and in other taxonomic groups (e.g. Barraclough *et al.* 1995; Owens *et al.* 1999; Arnquist *et al.* 2000). However, the relationship between floral symmetry and speciation remains untested (Waser 1998). I examine whether zygomorphy has the effect of increasing species richness in the angiosperm lineages where it occurs.

2. MATERIAL AND METHODS

(a) *Data collection*

I tested the null hypothesis that species numbers in zygomorphic clades were lower than, or equal to, the numbers in their actinomorphic sister clades. I considered symmetry only at the level of the corolla, ignoring the symmetry of the pistil and stamens. Although it is possible to have an actinomorphic corolla and zygomorphic gynoecium or androecium (e.g. *Hibiscus*), or vice versa (Neal *et al.* 1998), I limited the study to corolla morphology because it is the level of symmetry most likely to affect the pollination process (Stebbins 1974). Families in which corolla morphology was defined as zygomorphic were identified using Judd *et al.* (2002). If the information in that source was inadequate, I referred to Watson & Dallwitz (1992) or Mabberley (1997). Families described as having radially symmetrical, polysymmetric or regular corolla morphology were considered actinomorphic; those described as having bilaterally symmetrical, monosymmetric or bilabiate corolla morphology were considered zygomorphic. Only animal-pollinated families were considered.

(b) *Sister-group comparison*

Once I had exhausted the listed family descriptions I identified the phylogenetic relationships between these families using the angiosperm phylogeny created by Soltis *et al.* (2000). This is the most complete angiosperm tree currently available; it includes 75% of angiosperm families (Barraclough & Savolainen 2001). Sister-group analyses assume the inclusion of all extant species derived from each branch; hence the interpretation of the results assumes that there is no substantial bias towards actinomorphy in missing families. All the families I had identified as having primarily zygomorphic flowers were found on this tree. Upon identifying a zygomorphic clade I used the Soltis *et al.* (2000) tree to identify the actinomorphic sister clade. This process revealed that several of the zygomorphic families were in fact part of the same lineage. Eventually, 40 zygomorphic families yielded 19 sister group comparisons (figure 1).

Once the appropriate sister groups had been identified I used Mabberley (1997) to determine the number of species in each family. In cases where Mabberley (1997) disagreed with the taxonomic divisions in the Soltis *et al.* (2000) phylogeny, I used other sources (Watson & Dallwitz 1992 or Judd *et al.* 2002) to determine the number of species in the lineage. Occasionally, the zygomorphic families (e.g. Fabaceae) contained some actinomorphic members. Using methodology described in Farrel *et al.* (1991) and Heilbuth (2000), I reported the number

of species for the sister group as the total minus the number of actinomorphic species (figure 1). Similarly, in one case (Zingiberales) a group of taxa having wind-pollinated flowers (Poales) was removed from the zygomorphic sister-group total for the comparison. This procedure was conservative and could only bias the results against rejecting the null hypothesis. The reciprocal procedure (subtracting zygomorphic species from actinomorphic clades) was not performed; this also ensured that the test was conservative. While most sister groups represented independent comparisons, I included one sister pair (Polygalaceae–Surianaceae) that fell within the zygomorphic sister lineage of another pair (Fabaceae and its sister group). I controlled for any possible bias that this approach could have caused by subtracting the species from the Polygalaceae–Surianaceae comparison from the more inclusive sister group (leaving only the Fabaceae), thus assuring that one large group was not providing the basis for more than one positive comparison. However, removing this additional pair does not change the significance of the results reported below.

(c) *Statistical tests*

To determine whether there was a significant effect of the evolution of zygomorphy on the diversification rate of a lineage, I subtracted the number of species in the zygomorphic lineage from the number of species in the actinomorphic sister lineage. I tested whether there was a detectable trend in the direction of the differences using a one-tailed sign test and by testing whether the mean difference in species number between sister groups differed from zero using the non-parametric Wilcoxon signed-rank test. Results are reported as means ± one standard error.

3. RESULTS

In 15 out of 19 sister-group comparisons the lineage with zygomorphic flowers was more diverse than its sister group (table 1; figure 1: $p = 0.0096$, one-tailed sign test). Furthermore, the mean difference in species number between the sister groups was significantly greater than zero (table 1: $n = 19$, $p = 0.003$, Wilcoxon signed-rank test). The mean negative difference (actinomorphic clade contains more species) was 847.75 ± 758.17 and the mean positive difference (zygomorphic clade contains more species) was 3318.53 ± 1688.07.

4. DISCUSSION

The sister-group analysis leads to the rejection of the null hypothesis in favour of the alternative hypothesis that bilaterally symmetric (zygomorphic) clades are more species-rich than their radially symmetric (actinomorphic) sister clades.

This conclusion is consistent with field studies reporting an association between zygomorphy and species richness. In their study of 25 flowering plant communities, Ostler & Harper (1978) found that zygomorphy was correlated with increased plant diversity. Their explanation for this result is that in species-rich communities, zygomorphy should be favoured because it promotes increased fidelity between flowers of a given species and their pollinators.

It has been proposed that the evolution of zygomorphy will lead to increased speciation rates because it affects the precision of pollen transfer and hence the probability of

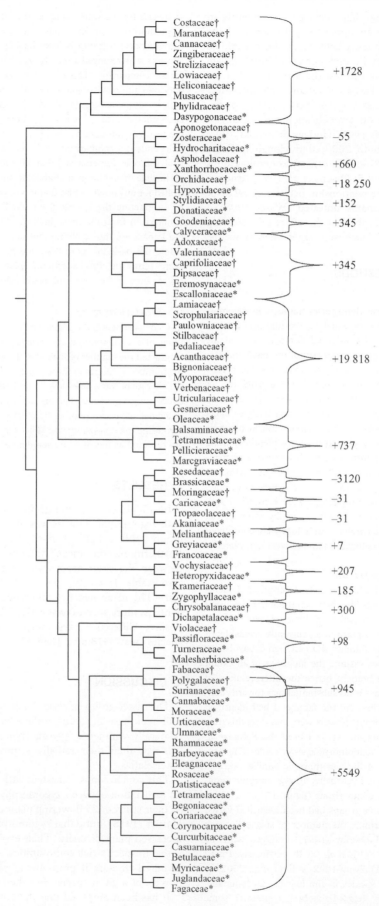

Figure 1. Phylogeny of zygomorphic angiosperm families and their sister taxa adapted from Soltis *et al.* (2000). Braces indicate the 19 sister-group comparisons. The number opposite each brace indicates the difference in species number between the two sister groups (zygomorphic species minus actinomorphic species); a dagger indicates a zygomorphic family; an asterisk indicates an actinomorphic family.

Table 1. Sister group comparisons for zygomorphic families. (Only animal-pollinated families were used. The numbers given have been corrected against this bias by removing actinomorphic members of the zygomorphic clade from the total (see text for details). The final column indicates the outcome of the sister-group comparison; + indicates that the zygomorphic clade had more species; − indicates that the actinomorphic clade had more species.)

zygomorphic family	number	sister group	number	+/−
Acanthaceae (3450) + Bignoniaceae (750) + Gesneriaceae (2900) + Lamiaceae (6700) + Myoporaceae (235) + Paulowniaceae (6) + Pedaliacae (85) + Scrophulariaceae (5100) + Stilbaceae (12) + Verbenaceae (950) + Utriculariaceae (245)	20433	Oleaceae	615	+
Adoxaceae (5) + Caprifoliaceae (420) + Dipsacaceae (290) + Valerianaceae (300)	1015	Eremosynaceae (150) + Escalloniaceae (1)	151	+
Aponogetonaceae	43	Hydrocharitaceae[b] (80) + Zosteraceae (18)	98	−
Asphodelaceae	750	Xanthorrhoeaceae	90	+
Balsaminaceae	850	Marcgraviaceae (108) + Pellicieraceae (1) + Tetrameristaceae (4)	113	+
Cannaceae[a] (8) + Costaceae (1100) + Heliconiaceae (80) + Lowiaceae (7) + Marantaceae[a] (535) + Musaceae (200) + Phylidraceae (6) + Streliziaceae (7) + Xyridaceae[b] (300) + Zingiberaceae (1100) − Bromeliaceae (1520)	1823	Dasypogonaceae	95	+
Chrysobalanaceae	460	Dichapetalaceae	160	+
Goodeniaceae[a]	400	Calyceraceae[a]	55	+
Krameriaceae	15	Zygophyllaceae	200	−
Fabaceae[b] (15315) − Polygalaceae[b] (950) − Surianaceae (5)	14360	Barbeyaceae (1) + Begoniaceae (900) + Betulaceae (110) + Cannabaceae (4) + Casuarniaceae (95) + Coriariaceae (5) + Corynocarpaceae (4) + Cucurbitaceae (775) + Datisticaceae (4) + Eleagnaceae (45) + Fagaceae (700) + Juglandaceae (59) + Moraceae (1100) + Myricaceae (55) + Rhamnaceae (900) + Rosaceae (2825) + Tetramelaceae (4) + Ulmaceae (175) + Urticaceae (1050)	8811	+
Melianthaceae	12	Francoaceae (2) + Greyiaceae (3)	5	+
Moringaceae	12	Caricaceae	43	−
Orchidaceae	18500	Hypoxidaceae	220	+
Polygalaceae[b]	950	Surianaceae	5	+
Resedaceae	80	Brassicaceae	3200	−
Stylidiaceae	154	Donatiaceae	2	+
Tropaeolaceae	89	Akaniaceae	1	+
Violaceae	800	Malesherbiaceae (27) + Passifloraceae (575) + Turneraceae (100)	702	+
Vochysiaceae[b]	210	Heteropyxidaceae	3	+
total				15+/4−

[a] Entire family displays secondary pollen presentation.
[b] Some members display secondary pollen presentation.

reproductive isolation arising among slight variants (Neal *et al.* 1998). If this were true, we would expect zygomorphy to be correlated with either specialist pollinators or the placement of pollen on specific parts of a pollinator's body. Additionally, I predict that other traits that require precise pollen transfer in order to have a selective advantage, such as lower pollen–ovule ratios, will be correlated with zygomorphy.

Indeed, an association between zygomorphy and pollination by specialist bees has been reported in several angiosperm taxa (Donoghue *et al.* 1998; Goldblatt *et al.* 2000). Specialist pollinators clearly have the potential to increase diversification rates. Bumble-bee pollinators may prefer zygomorphic to actinomorphic forms (Neal *et al.* 1998). In addition, bees may be inefficient pollinators of actinomorphic flowers (Cronk & Moller 1997). Moreover, reversals to actinomorphy may accompany a switch from specialist to generalist pollinators (Cronk & Moller 1997; Donoghue *et al.* 1998). There is also evidence suggesting that in some species with zygomorphic flowers, pollen placement is so precise that the same pollinator can visit multiple species and preserve reproductive isolation because the pollen is placed on different parts of the pollinator (Brantjes 1982, 1985). While further exploration is required, a correlation between zygomorphy and specialist pollinators further supports the hypothesis that higher species richness in zygomorphic lineages results from pollinator-mediated speciation.

If zygomorphy promotes reproductive isolation via improved placement of pollen, we would expect that the pollen–ovule ratio in species with zygomorphic flowers would evolve to be lower than in species with actinomorphic flowers. It has been demonstrated that the amount of pollen produced by a species (measured as the pollen–ovule ratio) is negatively correlated with the likelihood that the plant's pollen grains will reach a compatible stigma. For example, animal-pollinated plants have lower pollen–ovule ratios than wind-pollinated plants (Sharma *et al.* 1992), and plants that are obligately selfing (autogamous) have lower pollen–ovule ratios than those that obligately outcross (Cruden 1977). There is indeed some evidence that species with zygomorphic flowers have lower pollen–ovule ratios. For example, in the Orchidaceae, pollen is packaged into units known as pollinaria, which results in a pollen–ovule ratio that is several orders of magnitude smaller than in plants that lack these structures. The evolution of pollinaria has been directly attributed to the improved specificity accompanying the evolution of zygomorphy (Johnson & Edwards 2000). The pollinaria have been championed as a key innovation that allowed the rapid diversification of the orchid clade (Johnson & Edwards 2000). However, without a preceding adaptation to ensure highly specific pollination, pollinaria would be disadvantageous. In the Asterales, lineages that develop zygomorphy have often undergone a subsequent decrease in anther number (Endress 1998). While there are other possible explanations for this trend, it is an intriguing observation that deserves further exploration.

A potential problem with any sister-group analysis is that the examined trait (in this case zygomorphy) could be correlated with a different trait that drives diversification rather than be the actual cause of the diversification. This is an intrinsic problem with all correlative

studies. The presence of secondary pollen presentation, i.e. the presentation of pollen on floral structures other than the anther sacs (Yeo 1993), is also correlated with low pollen–ovule ratios (Cruden 2000), reportedly due to its ability to facilitate highly specific placement of pollen grains (Howell *et al.* 1993). Because of its purported role in improving pollination efficiency, secondary pollen presentation is another candidate trait that may play a role in angiosperm speciation. In addition, many families that display secondary pollen presentation also have zygomorphic flowers. Therefore, I repeated the sister-group comparison, excluding species or families that displayed secondary pollen presentation (table 1) to test whether secondary pollen presentation could have driven the association between zygomorphy and species richness. When these species are removed, only one comparison (Fabaceae and its sister lineage) is reversed, and the sign test remains significant ($p = 0.0155$). Because secondary pollen presentation is not strongly correlated with zygomorphy (table 1), it is unlikely to be driving the observed patterns of diversification. Secondary pollen presentation may also work in conjunction with zygomorphy in some families to ensure precise pollen placement (Yeo 1993).

A major weakness of a sister-group analysis is that it cannot distinguish whether differences between sister lineages in species richness are caused by more speciation events in one lineage or by more extinction events in the other. In the present case, however, there is no reason to expect that actinomorphy would increase extinction rates. Rather, actinomorphy may lead to lower extinction rates because of its association with generalist pollinators (Bond 1994; Johnson & Steiner 2000).

Finally, it is possible that zygomorphy affects speciation rates in a manner unrelated to its ability to promote reproductive isolation. For example, if reproductive assurance is greater in zygomorphic clades, they might be less susceptible to extinction. By contrast, zygomorphy may affect pollinator constancy such that competing species are more prone to extinction. The analysis presented here does not allow one to distinguish between these hypotheses.

In conclusion, I have argued that the correlation between zygomorphy and increased species richness in angiosperms is caused by the ability of this trait to promote reproductive isolation through improved precision of pollen placement and the tendency for specialist pollinators to be attracted to zygomorphic flowers. This study is distinctive in that it investigates a trait long suspected to be important in reproductive isolation and confirms a hypothesis central to evolutionary biology: traits that promote reproductive isolation are correlated with increased diversification rates.

The author thanks D. Ally for her assistance with data collection and methodology. The idea for this study was inspired through discussions with Q. Cronk, A. Mooers and the Vancouver Evolution Group (VEG). The methodology and manuscript were much improved by comments and ideas offered by A. Albert, J. Coyne, B. Davis, Q. Cronk, E. Elle, S. Iverson, R. Ree, H. Rundle, D. Schluter, J. Vamosi, J. Weir, M. Whitlock, J. Whitton and two anonymous reviewers. The author offers special thanks to S. Otto for her many insightful comments and suggestions. Funding for the project was provided through a Natural Sciences and Engineering Research Council of Canada (NSERC) PGS-B award to R.D.S. and a NSERC Discovery grant to S. Otto.

608   R. D. Sargent   *Floral symmetry and speciation*

REFERENCES

Arnqvist, G., Edvardsson, M., Friberg, U. & Nilsson, T. 2000 Sexual conflict promotes speciation in insects. *Proc. Natl Acad. Sci. USA* **97**, 10 460–10 464.

Barraclough, T. G. & Savolainen, V. 2001 Evolutionary rates and species diversity in flowering plants. *Evolution* **55**, 677–683.

Barraclough, T. G., Harvey, P. H. & Nee, S. 1995 Sexual selection and taxonomic diversity in Passerine birds. *Proc. R. Soc. Lond.* B **259**, 211–215.

Barraclough, T. G., Vogler, A. P. & Harvey, P. H. 1998 Revealing the factors that promote speciation. *Phil. Trans. R. Soc. Lond.* B **353**, 241–249. (DOI 10.1098/rstb.1998.0206.)

Bawa, K. S. 1995 Pollination, seed dispersal and diversification of angiosperms. *Trends Ecol. Evol.* **10**, 311–312.

Bond, W. J. 1994 Do mutualisms matter? Assessing the impact of pollinator and disperser disruption on plant extinction. *Phil. Trans. R. Soc. Lond.* B **344**, 83–90.

Brantjes, N. 1982 Pollen placement and reproductive isolation between two Brazilian *Polygala* species. *Pl. Syst. Evol.* **141**, 41–52.

Brantjes, N. 1985 Regulated pollen issue in *Isotoma*, Campanulaceae, and evolution of secondary pollen presentation. *Acta Bot. Neerl.* **52**, 213–222.

Cronk, Q. & Moller, M. 1997 Genetics of floral symmetry revealed. *Trends Ecol. Evol.* **12**, 85–86.

Cruden, R. W. 1977 Pollen–ovule ratios: a conservative indicator of breeding systems in flowering plants. *Evolution* **31**, 32–46.

Cruden, R. W. 2000 Pollen grains: why so many? *Pl. Syst. Evol.* **222**, 143–165.

Dodd, M. E., Silvertown, J. & Chase, M. W. 1999 Phylogenetic analysis of trait evolution and species diversity variation among angiosperm families. *Evolution* **53**, 732–744.

Donoghue, M. J., Ree, R. H. & Baum, D. A. 1998 Phylogeny and the evolution of flower symmetry in the Asteridae. *Trends Pl. Sci.* **3**, 311–317.

Endress, P. K. 1997 Evolutionary biology of flowers: prospects for the next century. In *Evolution and diversification of land plants* (ed. K. Iwatsuki & P. H. Raven), pp. 99–119. New York: Springer.

Endress, P. K. 1998 *Antirrhinum* and Asteridae—evolutionary changes of floral symmetry. *Symp. Ser. Soc. Exp. Biol.* **53**, 133–140.

Eriksson, O. & Bremer, B. 1992 Pollination systems, dispersal modes, life forms, and diversification rates in angiosperm families. *Evolution* **46**, 258–266.

Faegri, K. & van der Pijl, L. 1979 *The principles of pollination ecology*, 3rd edn. Toronto: Pergamon Press.

Farrell, B. D., Dussourd, D. E. & Mitter, C. 1991 Escalation of plant defense: do latex and resin canals spur plant diversification? *Am. Nat.* **138**, 881–900.

Futuyma, D. J. 1998 *Evolutionary biology*. Sunderland, MA: Sinauer Associates, Inc.

Galen, C. 1996 Rates of floral evolution: adaptation to bumblebee pollination in an alpine wildflower, *Polemonium viscosum*. *Evolution* **50**, 120–125.

Goldblatt, P., Manning, J. C. & Bernhardt, P. 2000 Adaptive radiation of pollination mechanisms in *Sparaxis* (Iridaceae: Ixioideae). *Adansonia* **22**, 57–70.

Grant, V. 1949 Pollination systems as isolating mechanisms in angiosperms. *Evolution* **3**, 82–97.

Grant, V. 1994 Modes and origins of mechanical and ethological isolation in angiosperms. *Proc. Natl Acad. Sci. USA* **91**, 3–10.

Grant, V. & Grant, K. A. 1965 *Flower pollination in the phlox family*. New York: Columbia University Press.

Heilbuth, J. C. 2000 Lower species richness in dioecious clades. *Am. Nat.* **156**, 221–241.

Hodges, S. A. & Arnold, M. L. 1994 Floral and ecological isolation between *Aquilegia formosa* and *Aquilegia pubescens*. *Proc. Natl Acad. Sci. USA* **91**, 2493–2496.

Hodges, S. A. & Arnold, M. L. 1995 Spurring plant diversification: are floral nectar spurs a key innovation? *Proc. R. Soc. Lond.* B **262**, 343–348.

Howell, G. J., Slater, A. T. & Knox, A. B. 1993 Secondary pollen presentation in angiosperms and its biological significance. *Aust. J. Bot.* **41**, 417–438.

Johnson, S. D. & Edwards, T. J. 2000 The structure and function of orchid pollinaria. *Pl. Syst. Evol.* **222**, 243–269.

Johnson, S. D. & Steiner, K. E. 2000 Generalization versus specialization in plant pollination systems. *Trends Ecol. Evol.* **15**, 140–143.

Judd, W. S., Campbell, C. S., Kellogg, E. A., Stevens, P. F. & Donoghue, M. J. 2002 *Plant systematics: a phylogenetic approach*. Sunderland, MA: Sinauer.

Kiester, A. R., Lande, R. & Schemske, D. W. 1984 Models of coevolution and speciation in plants and their pollinators. *Am. Nat.* **124**, 220–243.

Leppik, E. E. 1972 Origin and evolution of bilateral symmetry in flowers. *Evol. Biol.* **5**, 49–85.

Mabberley, D. J. 1997 *The plant book*. Cambridge University Press.

Neal, P. R., Dafni, A. & Giurfa, M. 1998 Floral symmetry and its role in plant–pollinator systems: terminology, distribution, and hypotheses. *A. Rev. Ecol. Syst.* **29**, 345–373.

Ostler, W. K. & Harper, K. T. 1978 Floral ecology in relation to plant species diversity in the Wasatch Mountains of Utah and Idaho. *Ecology* **59**, 848–861.

Owens, I. P. F., Bennett, P. M. & Harvey, P. H. 1999 Species richness among birds: body size, life history, sexual selection or ecology? *Proc. R. Soc. B Lond.* **266**, 933–940. (DOI 10.1098/rspb.1999.0726.)

Regal, P. J. 1977 Ecology and evolution of flowering plant dominance. *Science* **196**, 622–629.

Schemske, D. W. & Bradshaw, H. D. 1999 Pollinator preference and the evolution of floral traits in monkey flowers (*Mimulus*). *Proc. Natl Acad. Sci. USA* **96**, 11 910–11 915.

Schluter, D. 2001 Ecology and the origin of species. *Trends Ecol. Evol.* **16**, 372–380.

Sharma, N., Koul, P. & Koul, A. K. 1992 Genetic systems of six species of *Plantago* (Plantaginaceae). *Pl. Syst. Evol.* **181**, 1–9.

Soltis, D. E. (and 15 others) 2000 Angiosperm phylogeny inferred from 18S rDNA, *rbcL*, and *atpB* sequences. *Bot. J. Linn. Soc.* **133**, 381–461.

Stebbins, G. L. 1970 Adaptive radiation of reproductive characteristics in angiosperms I: pollination mechanisms. *A. Rev. Ecol. Syst.* **1**, 307–326.

Stebbins, G. L. 1974 *Flowering plants—evolution above the species level*. London: Edward Arnold.

Takhtajan, A. 1969 *Flowering plants: origin and dispersal*. Edinburgh: Oliver & Boyd.

Takhtajan, A. 1991 *Evolutionary trends in flowering plants*. New York: Columbia University Press.

Verdu, M. 2002 Age at maturity and diversification in woody angiosperms. *Evolution* **56**, 1352–1361.

Waser, N. 1998 Pollination, angiosperm speciation, and the nature of species boundaries. *Oikos* **81**, 198–201.

Waser, N. 2001 Pollinator behavior and plant speciation: looking beyond the 'ethological isolation' paradigm. In *Cognitive ecology of pollination* (ed. L. Chittka & J. D. Thomson), pp. 318–335. Cambridge University Press.

Watson, L. & Dallwitz, M. J. 1992 The families of flowering plants: descriptions, illustrations, identification and information retrieval, version, 14th December, 2000. http://biodiversity.uno.edu/delta/.

Yeo, P. F. 1993 *Secondary pollen presentation—form, function and evolution*. New York: Springer.

Guiding Questions for Reading This Article

A. About the Article

1. Name the journal and the year in which this study was published.

2. The author is affiliated with what university? In what country?

3. Funding for the project was provided through what agency?

4. Specialized vocabulary: Write a brief definition of each term.

   Described in the article text:

   actinomorphy

   zygomorphy

   Not defined in the article:

   androecium

   clade

   corolla

   gynoecium

   ovule

   phylogeny

B. About the Study

5. What is the adaptive explanation for the prominent evolutionary trend toward flowers with fused petals and reduced number of stamens and carpels?

6. Give two examples of specialized floral traits, and state the advantage of having such traits, from the plant's perspective.

7. State the genus names of two flowering plants for which experimental results suggest that reproductive isolation between two species is related to visitation by different pollinators.

8. Explain the proposed adaptive advantage of zygomorphy for improving specific pollen placement.

9. What was the source of the angiosperm phylogeny used for the sister-group comparison in this study, and why was that source chosen?

10. What is a null hypothesis, and what is the specific null hypothesis tested in this study?

11. In the process of science, proper hypothesis testing requires an experimental design with replication—that is, repeated independent tests of the same hypothesis with different data sets. How did this investigator design a study with replication?

12. The brackets at the left side of Figure 1 indicate phylogenetic groupings among zygomorphic angiosperm families, and the braces at the right of the figure indicate sister groups. (a) How many sister groups were compared in this study? (b) Look down the list and find the brace that includes the violet family, Violaceae. List the plant family names for that brace. (c) What do the brackets for the families in this brace indicate about the phylogenetic relationships among these families? (d) Which of the

Violaceae brace families is zygomorphic and which actinomorphic? (See figure legends for symbols.) (e) For the families in this brace, what is the number of zygomorphic species minus the number of actinomorphic species? (f) In this brace, which group has more species—the zygomorphic or the actinomorphic groups?

13. Do the results of the zygomorphic-actinomorphic comparison lead us to *reject* the null hypothesis or *fail to reject* the null hypothesis? Explain your answer.

C. General Conclusions and Extensions of the Work

14. How does the conclusion from this study compare to the results of field studies of zygomorphy and species richness in specific plant communities?

15. What relationship between zygomorphic floral structure and pollinator behavior has been observed in other research?

16. With any observation, an observed correlation does not always mean causation. (a) What is a potential problem (correlation ≠ causation) of the sister-group analysis of zygomorphy? (b) What *other* structural adaptation, besides zygomorphy, did the author suggest could be an alternative cause of the low pollen-ovule ratio that was observed? (c) How did the investigator address this alternative explanation? (d) What was the conclusion after the alternative explanation was eliminated?

17. What is a major limitation or weakness of the sister-group analysis for studying the relationship of zygomorphy and speciation?

18. What is the reasoning for predicting that plant species with zygomorphic flowers would evolve a *lower* ratio of pollen produced to number of ovules produced? Are there any plants that show such a low pollen-ovule ratio?

19. In concluding, the author states that "the correlation between zygomorphy and increased species richness in angiosperms is caused by the ability of this trait to promote reproductive isolation through improved precision of pollen placement and the tendency for specialist pollinators to be attracted to zygomorphic flowers." Do you think that her study provides proof of (a) that correlation and (b) that causation?

20. Imagine that you were working with this researcher. What could be a possible follow-up test that extends this work?

PART B

READING PRIMARY LITERATURE:
A PRACTICAL GUIDE TO EVALUATING RESEARCH
ARTICLES IN BIOLOGY

CHRISTOPHER M. GILLEN

Contents

Section 1: Introduction to the Booklet

Primary research articles are an excellent resource for students of biology. Reading papers opens a doorway into the world of scientific research. As you begin reading articles, you will be challenged to think critically, apply your knowledge, and use the scientific method. The purpose of this booklet is to guide you through this process. It describes the tools and strategies you'll need to begin reading articles.

WHAT ARE PRIMARY RESEARCH ARTICLES?

Primary research articles, also called **research papers** or **primary literature**, are the official documents that scientists use to communicate their research to each other. Research papers describe original findings, including methodology and results. In contrast, documents that synthesize, summarize, or evaluate primary literature are termed **secondary** sources. Common secondary sources include magazine articles, dictionaries, textbooks, and websites. Magazines written for the general public, such as *Discover* and *Scientific American*, publish only secondary articles. Some publications, such as the scientific journals *Science* and *Nature*, publish a mixture of primary research papers and secondary reports. Primary papers are always written by scientists, while secondary reports may be written by scientists, journalists, or others. In practice, you'll need to look at each source carefully to determine if it is primary or secondary. Primary sources include a detailed description of the methods and results. They are the first report of original research findings and are written by the scientists who performed the work. This book focuses on primary research articles.

You've probably spent hours reading science textbooks. They describe accepted scientific facts and concepts, and they are written towards a student audience. In contrast, research articles address areas of emerging knowledge or controversy, and they are written with professional scientists as the intended audience. Research articles represent science in process, while textbooks describe the outcome of that process. Reading research articles requires different strategies than reading textbooks, and you will encounter new challenges as you begin reading them. This booklet will give you the tools to overcome the challenges.

WHY READ THE PRIMARY LITERATURE?

Doing science is a powerful way to learn it. In your laboratory classes, you begin participating in science by performing key experiments and techniques. Reading research articles is another way to become involved in the scientific process, because evaluating research articles is a central activity of practicing scientists. Think of research papers as a doorway into the scientific world. They present the chance to apply material that you've learned through reading textbooks and listening to lectures. They challenge you to think along with scientists as they tackle research problems. They serve as examples for your own scientific writing. And they encourage you to critically analyze new scientific ideas.

Try to approach reading research articles as a challenge rather than a chore. While the primary literature may appear dry, under the surface you will find drama and controversy. Research scientists constantly face mysteries because they work at the edge of our knowledge. So research articles can be like detective novels, in which scientists carefully gather the clues and evidence needed to solve scientific problems. Further, like a good detective novel, the course of science often has entertaining twists and surprises. Finally, the research literature reflects upon the scientists themselves. It documents how their tools, priorities, and practices change over time and how they cooperate and compete with each other. Recognize the personal side of science; it will add useful context to the research articles you read.

Learning to read research articles has obvious benefits if you plan to pursue a research career in biology. But the brief exposure to the primary literature afforded by this booklet will be beneficial even if you never again read a primary literature article. As a child, I took up the trumpet and learned to play the "Star Wars Theme" loud enough to drive our Irish Setter under the kitchen table. However, I never became especially skilled, and I can't play a note today. Yet my battles with the trumpet continue to enrich my adult life by enhancing my appreciation and understanding of music. In the same way, learning to read primary literature opens the door towards comprehending and evaluating all sorts of scientific information. Learn to read research articles and you will become a more confident and effective judge of scientific information, a remarkably useful skill in today's world.

HOW TO USE THIS BOOKLET

Learning to read primary research articles is a lot like learning other demanding activities. For example, imagine that you've been handed a fishing rod and a tackle box. Without any advice, trying to catch fish using these tools will be hard. Some tips and advice from an expert will greatly ease your way. But instruction alone won't make you a competent angler. You'll need to go stand by the water and practice baiting a hook, casting a line, and landing a fish. So it is with reading research articles; both *instruction* and *practice* are needed.

This booklet is the instruction manual you'll need to quickly become a talented reader of research articles. The sections cover how to find articles, the structure of research articles, and the four main parts of an article: the Introduction, Materials and Methods, Results, and Discussion. In addition to studying this booklet, you'll also need to struggle with research articles yourself. The end-of-section exercises will prompt that practice; they ask you to critically read a paper by applying the section's advice. Depending on your course and instructor, you might choose your own article using the instructions in section 2 or your instructor might assign one.

As you begin your journey into the primary literature of biology, there will inevitably be obstacles and frustrations. But I hope you will also find satisfaction in overcoming the challenges and participating in the process of science.

Section 2: Finding Research Articles

The major journals in biomedicine now publish online. More than a million scientific articles are available free on the Internet, and many more can be accessed for a fee. Along with this availability comes challenges. Scientists need to find relevant information amidst a mountain of data and then intelligently analyze the information. This section describes the first step: locating and accessing research articles. The details will be specific to your institution and subdiscipline, so we will focus on general strategies and free online databases. You should also learn about the specific resources available on your campus.

The strategies you employ will depend on the purpose of your search. Are you trying to develop a topic for a research paper? Scanning the table of contents of a prestigious journal might give you some ideas about current hot research areas. You might also look at secondary sources such as textbooks and magazines for ideas. If you've already selected a general topic area, but need to find articles within that topic, then you might decide to search databases of research articles. Or perhaps you've already found an interesting article and now need to read some related articles to put it in context. In this case, you might look at the sources it cites and also search databases.

WINDOW-SHOPPING: BROWSING JOURNALS

Suppose you wish to browse journals to develop topic ideas. Where can you find journals? One place is your school's library, which has hard copies of many journals. Having the paper journal in your hands can be particularly useful when you are browsing for interesting topics. Of course, many journals are also available online. If you already know which journal you want to browse, a simple strategy is to locate its homepage with an Internet search engine. Online journals have straightforward interfaces that allow you to browse or search for articles.

Another way to access online journals is through a website that catalogs and organizes them. For example, you will find many important journals in biomedicine at the Highwire Press site hosted by Stanford University Libraries (www.highwire.org). Excellent instructions for using the site are available on its homepage. You can see a list of journals, choose one that matches your interests, and browse the tables of contents of recent issues. For many journals, access to the most recently published articles requires a subscription, with free access available about a year after publication. A useful feature of the Highwire site is a listing of journals with free online access.

Almost 1000 scientific journals are hosted on Highwire. How can you decide which journals are the most relevant to your topic? Which are the most reliable? Which are the most prestigious? One strategy is to look in the references section of your class textbook to see which journals are cited; these are likely to be relevant to your course and highly regarded. As you begin to find research articles, you can also check their References Cited sections; again, the journals that are regularly cited are the most useful and prestigious. Journal rankings also exist. One widely used measure, the **impact factor**, measures how many times the articles published in a journal are cited by other articles. Finally, you can ask your course instructor or librarian. Widely read and cited journals that publish research in biology include: *Science, Nature, Cell, Development, Genetics, Proceedings of the National Academies of Science (PNAS), Journal of Biological Chemistry, Journal of Clinical Investigation, Journal of Experimental Biology, American Journal of Physiology, Neuron, Journal of Bacteriology,* and *Public Library of Science (PLoS) Biology.* This is a partial list; there are many other outstanding journals.

Another useful website is BioMed Central (www.biomedcentral.com), which publishes more than 150 journals, including general titles such as *BMC Biology* and *Journal of Biology* and specialized titles such as *BMC Physiology* and *BMC Microbiology.* Articles published in BioMed Central are **open-access**, meaning that they can be freely accessed and distributed. You can search all of the BioMed Central journals or browse particular titles.

Online articles are usually available in two formats: full-text and PDF. PDF files, which are viewable using the Adobe Reader®, faithfully represent the article as it appears in the print journal. Access the PDF version if you wish to print the article. In contrast, the full-text version of an article does not attempt to represent the actual printed article; instead full-text versions use html format and therefore may include special features not available in the printed article. For example, links may enable you to navigate easily throughout the article, to access supporting documents such as supplemental data, and to go to full-text versions of other articles.

Suppose you have chosen to work with a particular article. The html version of your article can be an entry point into an interconnected web of references. You may find links to other articles, including those by the same authors, those that your article cites, and those that cite your article. In contrast to the printed version of an article, the html version can be updated, so that links can be included to sources that were published after the original paper. For example, you can't identify papers that refer to your article by looking at the printed version, but the html version may contain this information. Locating such sources can be valuable since they may offer commentary on your article. The *Science Citation Index*, which is available at many institutions, is another tool that enables you to identify sources that cite a particular article.

NEEDLE IN A HAYSTACK: SEARCHING DATABASES

Suppose you have developed a research topic and need to find relevant primary sources. Or maybe you have found a single journal article and need to find additional articles on the same topic. In both cases, searching a research article database will be an effective approach. Several databases may be available on your campus; choose the one best suited to your topic. Although details will vary slightly from one to the next, the general principles will be the same, and most databases contain helpful instructions to get you started.

As an example, we'll discuss the freely available PUBMED site, which is part of an interconnected collection of databases operated by the National Center for Biotechnology Information (NCBI, www.ncbi.nlm.nih.gov). PUBMED allows users to search articles in almost 5,000 journals in medicine and related fields. In total, over 15 million citations are included. Given the huge number of citations in PUBMED, separating useful articles from all of the rest can be tricky. Good searches will be *comprehensive*, meaning they will return all or most of the articles on a particular topic, and will also be well *focused*, meaning they will return a reasonable number of articles with few irrelevant articles.

The choice of **search terms** is obviously crucial to an effective search. You can find search terms on the first page of research articles; look for a list of terms below the Abstract. You might also pick up useful terms in secondary sources or by browsing article titles. Scientists often use specialized words with precise meanings, so it also helps to know some of the scientific vocabulary. For example, while the general public says "heart attack," a scientist may say "sudden cardiac death," "myocardial infarction," "heart failure," "cardiac ischemia," or "ventricular fibrillation," with each of these phrases having a somewhat different meaning. Searches using different terms will return different sets of articles. Knowing exactly which terms to use can be difficult to determine, and some trial and error may be needed. Don't be afraid to start searching with a term that might be imperfect. Searches are free and fast, and better search terms can often be gleaned from the citations returned by an initial search.

Because scientists may use various terms for the same concept, many databases include a **controlled subject vocabulary**. In PUBMED, these terms are called the Medical Subject Headings (MeSH). You can find MeSH words by searching the MeSH database on the NCBI site. For example, searching for "cardiac ischemia" returns two MeSH terms: "myocardial ischemia" and "coronary arteriosclerosis." The term that best matches your interest can be used as a search term, leading to a comprehensive and well-focused search.

Once you've identified some useful search terms, a good strategy is to start with general searches and then try to focus. Preliminary searches may yield an unmanageable number of articles. For example, PUBMED returns about 250,000 articles for the search term "hypertension" and 200,000 articles for "diet." Combining terms using AND will focus the search. About 10,000 articles are returned for "diet AND hypertension," a lot less than either term alone but still probably too many to sift through. The search can be further narrowed using the LIMIT function, which enables you to select features such as publication date range, language, subject (animal or human), subject age, and online availability of the full-text article. You can also restrict your search to only article titles or abstracts to narrow the search further.

Using the operator OR between terms will return all the citations that contain either term and thus can be useful for generating a comprehensive citation list. For example, if you wish to make a complete search about athletics, you might combine "athletics" with "sports" using OR. Searching "athletics OR sports" will return more articles than using either term alone. The NOT operator is also handy. Placing NOT before a search term eliminates all articles containing the term, thereby reducing the number of articles returned. Suppose you find that a major portion of the literature on sports and athletics deals with badminton, a sport that does not interest you. NOT could eliminate articles about badminton, thus returning a more reasonable total number of articles.

Searches that return between 10 and 100 articles are good starting points. You can scan the titles for articles that match your interest. Both Highwire and PUBMED mark free access articles with special icons, so you can easily identify them. However, be careful not to let free availability be a major search criterion. You may have access to many articles not listed in these databases as free. Your library may subscribe to the journal either in paper or online, or you may be able to obtain the article through an interlibrary loan. Check with a librarian at your institution for more information.

SECTION 2 EXERCISES

These exercises will help you locate and access research articles:

1. Visit your library or access an online database. Identify several journals in your area of study that publish primary research articles. Browse the table of contents of these journals and scan some of the articles. List two or three differences among the journals you've found.

2. Locate the journals *Nature* (www.nature.com) and *Science* (www .sciencemag.org), again either online or at the library. Browse through an issue of one of these journals and identify primary and secondary articles. Write down the titles, authors, journal, year, volume, and page numbers of three primary and three secondary articles.

3. Develop a list of three research topics that interest you, using your textbook, popular magazine articles, Internet news sites, and other secondary sources.

4. Develop a list of three current research topics by browsing through the table of contents of recent journal issues. Look for topics that are explored in several recent articles.

5. Choose one of the research topics from question three or four and develop a list of five search terms that could be used to search for research articles on the topic.

6. Using an appropriate database, identify five primary research articles on your topic.

7. Pick one of the articles from question six. Develop a list of at least five research articles that would help you understand the chosen article. Look for articles cited by the chosen article, articles that cite it, and other articles by the same authors. Also, search appropriate databases to find additional related articles.

Section 3: The Anatomy of a Paper

Most biology research papers contain the following sections: Citation, Abstract, Introduction, Materials and Methods, Results, Discussion, Acknowledgments, and References. In this section, we first consider the ways that you can use the format of papers to read efficiently and critically. We then discuss the Citation, Abstract, Acknowledgments and References sections. Subsequent sections will be devoted to the Introduction, Materials and Methods, Results, and Discussion.

FORMATTING MATTERS

Research articles share a common **format**; material is presented in discrete sections arranged in a particular order. For instance, the Materials and Methods section, which describes exactly how the work was done, usually precedes the Results section, which portrays the data. This formatting is not only convenient for readers and writers; it also enables research to be judged according to the standards of the scientific community.

The format of research papers can help you maintain a critical stance. Because each aspect of the study is described in a separate section, you can independently assess different aspects of a study. For example, original data are presented entirely in the Results, while interpretation of data is mainly confined to the Discussion. You can exploit this separation by closely analyzing the data in the Results before considering interpretations in the Discussion.

The format of papers also allows you to quickly access information. To make the most of this, try to approach a paper with specific objectives. You may want to use the Introduction as background information about a new topic. You may focus on the Materials and Methods if you wish to find techniques to use in your own research project. You may want to compare the study's Results with those of another study. Or you may be interested in how a famous scientist synthesizes new research in the Discussion. Make yourself familiar with the format of papers; it will allow you to go directly to the desired information.

Journals vary in their formatting. Sometimes the Materials and Methods are at the end of the article rather than following the Introduction. Sometimes the Results and Discussion are combined into a single section. Sometimes, most notably in the very influential *Nature* and *Science*, the text is not divided into separate sections. However, if you are familiar with the standard format of papers, you can adjust to these modified formats. Look for writing that corresponds to each of the sections of a standard article.

The format of papers is not only helpful to readers; it may also impact the scientific process. Scientists need to communicate their findings effectively in research articles, so they may anticipate the paper they plan to write while they are collecting data. In this way, the demands placed on authors by the format of papers can influence the way they conduct their studies. This influence is overwhelmingly positive, because the format of papers is consistent with the methodological standards of the scientific community. If you are conducting independent research, you may find it useful to think about how you will present your work before you perform your studies. Will you be able to make a convincing case to other scientists?

CITATION

Basic **citation** information is given at the top of an article's first page: the title, authors, institutional affiliations, journal, volume, pages, and publication date (Figure 1). Don't skip this information; it can be surprisingly useful. The journal name, volume, and page numbers form a unique "address" and can be used to identify a particular article. You can often find the journal's **publisher** on the first page. Some journals are published by scientific societies, for example the American Physiological Society or the American Society of Microbiologists. Others are produced by commercial publishers. While journals produced by societies and companies can both be reliable, it helps to know what organization is behind the journal. Also scrutinize the **title**, which usually includes the species studied, the experimental approach, and perhaps a brief indication of the results obtained. Reading article

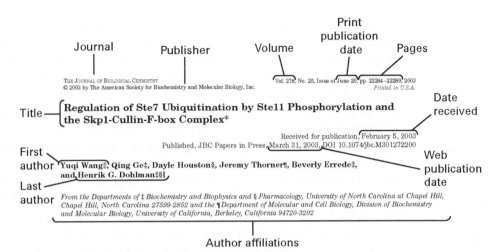

FIGURE 1. The header of a scientific paper. In this example, a Web publication date is given rather than a paper acceptance date. Adapted with permission from the American Society for Biochemistry and Molecular Biology.

titles carefully can save you time; you can often decide whether you wish to read any further.

Authors: The Research Team

The stereotype of the lonely scientist working in the solitary confinement of a laboratory stands in direct contrast to the collaborative, interdisciplinary world of modern science. Most modern research articles have multiple authors, reflecting this collaborative model. Let's look at the personnel in a typical research group and explore how they are represented in an author list.

Scientists work in diverse research groups that ordinarily include people of different ages, backgrounds, and educations. The boss is the **principal investigator (PI)**, an established scientist who usually has a PhD or MD. PIs determine research priorities, write grants, hire personnel, present findings at conferences, convene lab meetings, and revise manuscripts. They are ultimately responsible for the work done in their lab. Research groups contain other scientists who have finished their education. **Postdoctoral fellows** ("post-docs" for short) have recently received a PhD or MD, but have not yet established their own research program. They are experienced, free from other responsibilities, highly motivated, and often the most productive members of a research group. Groups may also include **senior research scientists** who are well past the post-doc portion of their career and have a long-term position. Finally, **collaborators** may join a research group for a particular project. Collaborators may come from across the hall or from across the globe, and many laboratory groups are distinctly multinational.

Graduate students are important members of many research groups. They each have a specific project which will culminate in a written thesis and an oral defense. Their research both contributes to science and fulfills a requirement of their degree programs. Master's programs usually last one or two years while PhD programs may last between four and six years. **Undergraduate** students sometimes work in labs during the summer or part time during the school year.

Technicians are employed by a lab and are not working towards a degree. Some are permanent members of a research group while others are recent college graduates who plan to work for a year or two before applying to graduate or professional school. Technicians are usually responsible for aspects of the day-to-day work of the lab. They perform experiments, order supplies, analyze data, and train students. An experienced technician can be a central member of a lab group, keeping the lab running smoothly and providing long-term continuity to the lab's workings.

The sequence of authors contains useful information (see Figure 1). The **first author** is usually the scientist who did the most work on a project, often a senior scientist, post-doc, or graduate student. The PI is often listed as the **last author** and is also sometimes called the **senior author**. The first and last authors get a major portion of the credit for the study. **Middle authors** may have contributed in any number of ways, for example, by providing technical help or by reading and editing the manuscript. They get considerably less credit than first and last authors. If you wish to track down similar studies, focus your attention on the first and last authors, who are most likely to have done additional work on the topic.

Who is included as an author? Scientific ethics require that only those who have made an intellectual contribution to the work be listed. It is unethical to omit someone who has significantly contributed to a project or to include someone who has not had a substantive role. In the end, each author is responsible for the content of the paper. When technicians or undergraduates have made an intellectual contribution to the project, they are listed as authors. Otherwise, they might be thanked in the Acknowledgments.

Consider the institutional affiliations of the authors. Are all of the authors from the same institution or is this a multi-institutional collaboration? What sorts of institutions are involved? Major research universities? Drug manufacturers? Liberal arts colleges? Are the authors associated with basic science departments (i.e., Cell Biology, Biochemistry, Physiology) or are they associated with clinical departments (i.e., Internal Medicine, Nephrology, Oncology)? The institutional affiliations can tell you about the authors' perspectives.

Publication Dates: Peer Review
On the first page of an article, you may find several dates, including the date the article was received by the journal, the date it was accepted for publication, and the print publication date (see Figure 1). Articles are sometimes posted on the Internet shortly after acceptance, and in such cases the date of first Internet publi-

cation may be given rather than the acceptance date. Why are there delays between receipt, acceptance, and publication? What happens in these interims? Let's look at the process of drafting, submitting, and revising a manuscript.

One scientist often writes a first draft of a research paper. After this initial effort, however, the process is usually distinctly collaborative. The draft will typically be read by several members of a research group, often resulting in substantial revisions. The manuscript may also be read by scientists outside the immediate research group, who can provide an objective perspective. After this initial "friendly" review, a manuscript may be ready to be sent to a journal. Choosing the appropriate journal is an important decision for authors. Authors seek to publish their work in prestigious, widely read journals. However, the review process can be lengthy, so it is risky to send an article to a top-notch journal where acceptance may be uncertain.

Submitted articles are first considered by an editor who assesses whether the manuscript is within the scope of the journal. If so, the editor will send it out to two or more anonymous reviewers. This process is called **peer review,** because the reviewers are the authors' peers: other research scientists with expertise in the study's topic. They critically read the manuscript and inform the editor of its suitability for publication. They may also be asked to rate the priority or importance of the manuscript. Reviewers also write a set of comments for the authors; these will be essential to the authors if they are asked to revise and resubmit. The final decision about publication is in the hands of the editor. The editor may simply reject the manuscript, accept it without revisions, ask for minor revisions, or require major issues to be addressed before resubmission.

Although you will not find any direct evidence of the peer review process in most papers, remember that each primary research article has been evaluated by other scientists before publication. One clue to the peer review process is the delay between submission and acceptance (or Internet publication) dates. If the article was accepted shortly after submission, it could mean that the reviewers examined it carefully and found it flawless. Or they might have looked at it quickly and missed an important issue. If there was a long delay, the authors might have been asked to make substantial revisions, perhaps even to supply additional data. On the other hand, a reviewer might simply have been late in submitting comments. Take note of the length of the delay between submission and acceptance, but don't put too much emphasis on it.

Once a manuscript is accepted, it will still be some time before the article is finally printed. Before publication in print, authors receive a formatted version of the article. They can respond to editorial changes, check for small errors, and answer questions from the publisher. Because the review and publication process can be lengthy, scientists find it helpful when the submission, acceptance, and publication dates are printed on the paper. Scientists get credit by publishing *original* findings. When multiple groups publish on the same topic, submission and acceptance dates can help resolve disagreements about who published first.

You can use these dates to help understand what the authors knew when they submitted an article. For example, an article submitted in October 2002 and published August 2003 could not reasonably be expected to consider information that was published in June 2003.

Peer review is a good way of evaluating the scientific value of papers before they are published. Errors are corrected, interpretations are refined, and explanations are deepened. Work that fails to meet basic standards isn't published. But peer review isn't perfect. Mistakes sometimes slip past reviewers, so don't be surprised if you find something in a paper that might be incorrect. Also, peer review isn't designed to uncover scientific fraud. In the rare cases where scientists intentionally misrepresent their work, peer reviewers are usually not in a position to detect the fraud. The system thus relies on the honesty of authors, and breaching this trust is considered to be an ethical lapse of the highest order. Fortunately, fraud is often exposed through other mechanisms, either by whistleblowers who have firsthand knowledge of the misconduct or as a consequence of other scientists failing to replicate key findings.

ABSTRACT

Abstracts are succinct summaries of research papers. They usually include statements of the study's purpose, experimental approach, key results, and conclusions. Each of the sections of a scientific paper is condensed into a few sentences in the Abstract. They are typically limited to less than 250 words, so each word must be carefully chosen.

Apart from the title, Abstracts are the most widely distributed portion of papers. They are freely available in online databases. Take the time to read the Abstract closely. It will introduce you to the study's core methods, findings, and conclusions, and help focus your further reading. Some Abstracts will be difficult to read. When an entire research study is condensed into a single paragraph, the writing must become quite dense. In some cases, it might be best to skim the Abstract and proceed to the other sections, where you will find more explanation. Don't quit reading an article just because the Abstract is too difficult. Give the rest of the paper a chance. Finally, resist the temptation to use the Abstract as a substitute for reading the full article. You won't find enough detail to make an informed judgment. Scientists don't cite an article solely based on the Abstract.

When scientists attend meetings, they frequently present their findings in short oral sessions or as posters. Such presentations are often accompanied by an Abstract, which is distributed to meeting attendees and sometimes published in a special journal issue. These Abstracts are similar to those that accompany a full-text scientific paper. But they are not supported by full-text papers, and they do not always receive critical peer review by other scientists. Thus, meeting abstracts may be less reliable documents than Abstracts of full-length papers. Much of the work presented at meetings will eventually be incorporated into peer-reviewed research articles, and meetings are an important way for scientists to get feedback on their work prior to publication.

ACKNOWLEDGMENTS

In the **Acknowledgments**, authors thank the people or institutions that have contributed to the work. Scientists acknowledge those who have given valuable feedback, either by reading a draft of the manuscript or by commenting on a preliminary presentation of the work. You can sometimes get a good sense for the pre-publication feedback that scientists received by seeing who they thank in the Acknowledgments. Authors may also acknowledge technical assistance and gifts of equipment, supplies, or reagents.

Research funding sources are noted in the Acknowledgments. Numerous sources of money are available to scientists, including the home institution, government agencies such as the National Science Foundation and the National Institutes of Health, industry groups such as pharmaceutical companies, and private research foundations such as the American Heart Association. Scientists apply for funding through **grant applications** that describe their past accomplishments and the proposed studies. These applications are evaluated by panels of scientists. Check the Acknowledgments to see how a study was funded. Since competition for grant monies can be extraordinarily competitive, funded projects have passed a rigorous peer review. Consider also whether the funding source could bias outcomes. Would you feel differently about a study funded by a pharmaceutical company compared to one funded by the government? Perhaps it would depend on the purpose of the research? Finally, post-docs and graduate students may receive fellowships to fund their studies. When these are noted in the Acknowledgments, it can help you identify which authors are students and post-docs.

Some journals ask authors to disclose possible conflicts of interest, usually in the Acknowledgments or in a separate statement at the end of the article. For example, scientists might have a financial interest in a company whose business is related to the study. Check to see if any of the authors have a direct financial stake in the outcome of the study; this might influence the way you evaluate the article.

REFERENCES

The **References** section is sometimes called the **References Cited** or **Literature Cited**. It includes only those articles cited in the text of the paper. By citing other studies, scientists acknowledge the work of others and position their own work within a larger body of scientific literature.

The References can be an excellent source for finding further reading on a topic. You may wish to analyze the kind of sources used in a paper. Have the authors previously published on this topic? Do they mainly cite their own works? Have the authors considered other recent studies? Have they read the older, classic literature on a topic? Have they considered the work of scientists outside their immediate field? Answers to some of these questions may be found with a quick look at the References, and if you wish to pursue the matter further you can read some of the cited articles.

THE EXTENDED RESEARCH TEAM

We have seen in this section that science is very much a team activity. The research team extends even beyond the list of authors to include other scientists from within and outside a research group. Members of a research team who are not authors often give feedback by reading drafts of the manuscript. They may also share opinions during **lab meetings**, which are regular meetings of research groups. Results and interpretations are usually given a rigorous challenge at these meetings; scientists evaluate their own findings critically before sharing them with other research groups.

Scientists outside the research group contribute to the work by reviewing manuscripts, assessing grant proposals, and offering technical help. They also comment on conference presentations and **research seminars**. Academic departments regularly invite speakers, often from outside their own institution, to present seminars. Speakers present their new findings and often receive questions and useful feedback. The audience learns about new research, often before it is published. Keep these inputs from other scientists in mind when reading a paper. By the time a paper has been published, the opinions of many scientists have been incorporated.

SECTION 3 EXERCISES

Using a research article as an example, complete the following exercises:

1. In one or two sentences, restate the title of the paper in a way that would be understandable to a member of the general public without a scientific background.

2. Who are the authors of the paper? What kind of institutions are they from? What kind of departments are they in? Do the institutional or departmental affiliations of the authors offer any insight into their perspective or possible biases? Can you identify the PI? Are any of the authors students?

3. When was the paper published? How long were the delays between submission and acceptance and publication?

4. Read the paper's Abstract. Summarize the main point of the study in two or three sentences.

5. Can you determine how the study was funded? If so, does the source of funding influence your opinion of the work?

6. Were any other scientists consulted in this project? Did the authors get feedback from other scientists prior to publication?

7. Examine the References. Do the authors cite themselves? Are some other authors cited frequently? Are recent works cited? Do the authors cite any older papers to provide an historical perspective? Do you see any sources that might help you understand the paper better?

Section 4: The Introduction

The Introduction contains background information and a description of the study's purpose. Authors describe a research problem, explain prior work, and indicate where controversy exists. They describe why their work is important and how it seeks to extend knowledge. You can use the Introduction to learn about previous studies in a research field and to understand the study's purpose.

UNDERSTANDING THE JARGON

Suppose you encounter an Introduction that is littered with unfamiliar technical language. It helps to know that scientists have valid reasons for employing specialized terminology. Scientific words or phrases can condense a large body of shared knowledge. For example, when cell biologists say "tyrosine kinase," they mean "a member of a class of enzymes that catalyzes the addition of a phosphate group to the amino acid tyrosine in certain proteins thereby affecting their function." Writing this every time would be cumbersome, so the technical language serves an important role. Taking the time to learn the specialized vocabulary may be time-consuming, but it's a necessity if you hope to appreciate the paper's scientific concepts.

To comprehend a paper's specialized terminology you may need to consult additional sources. Secondary sources, including textbooks, review articles, websites, and dictionaries, are useful tools for understanding scientific terminology and concepts. Some secondary sources are written specifically for nonscientists, and are easier to understand than research articles. However, because secondary sources are so varied, it's necessary to carefully evaluate the credibility of each source. Also, secondary sources must never substitute for careful reading of primary research papers. Scientists always consult primary sources on questions of central importance to their own work.

To master the terminology in a paper, note the most commonly used technical terms in the Introduction and then find definitions for them, using a biological, scientific, or medical dictionary. Such dictionaries can be found online and in the reference section of libraries. While looking for challenging terminology, pay particular attention to abbreviations and acronyms. These are often used extensively and can make an article seem incomprehensible. Abbreviations are sometimes defined in a list on the article's first page. Take the time to become familiar with each abbreviation.

Defining terms is a good start, but to fully understand the concept behind each term you may need to go beyond dictionaries and consult other sources. Books, including textbooks, can be a good tool. You probably know that as knowledge in biology has grown, textbooks have lengthened accordingly. Fortunately, using texts to understand the primary literature does not require a cover-to-cover read. Instead, you can identify topics in the table of contents or index, and read only the relevant sections. Encyclopedias and magazine articles can be used in similar fashion.

Review articles offer more specialized and focused background reading. They are often found in the same journals that publish primary research articles, but their focus is summarizing and synthesizing the findings of many studies rather than presenting new results. Reviews are written as communications to practicing scientists, and thus may be more difficult to understand than textbooks or magazine articles. However, they can provide an authoritative overview of a research field and usually have comprehensive references lists that can be a good source of further reading.

Information on the Internet can be both current and convenient to access. However, the quality of Internet sites is variable; some are trustworthy, others are not. Thus, evaluating the reliability of Internet sites is essential. Here are six elements that should be considered when evaluating websites:

1. Author. Who is the source of the information? Is there a single author or an organization? What are the author's qualifications? Is the site affiliated with the government, an educational institution, a private foundation, or a company?

2. Scope. What is the intended audience of the site? What body of information is covered? How does the scope relate to the author's expertise?

3. Timeliness. When was the information posted? How often is the site updated?

4. Presentation. Are there misspellings or grammatical errors? Are there broken links?

5. Mission. Does the site have an obvious agenda? Is there any obvious bias?

6. Review. Has information on the site been peer reviewed? Is there a mechanism for comments, feedback, or criticism?

Among the most highly reliable websites are those developed by universities and colleges, government health and science agencies, professional scientific societies, and private research foundations. However, even websites hosted by reputable organizations should be critically evaluated. In fact, it is best to make a habit of assessing the reliability of *every* secondary source; with minor modification, the previous guidelines can be used for printed sources.

OBSERVATIONS, EXPLANATIONS, EXPERIMENTS

Scientific studies are rooted in previous work, yet seek to expand the boundaries of existing knowledge. A key aspect of the Introduction is describing the study's purpose within the context of prior studies. Appreciating this function of the Introduction requires an understanding of the different sorts of activities that make up scientific methodology. Scientists make observations, propose explanations, and test explanations. Although these processes are interconnected, studies do not necessarily address all three of them. In reading the Introduction, you should identify the main purpose of the study and ask questions appropriate to the purpose.

A main goal of biological inquiry is to develop accurate explanations of the natural world. Scientists explore areas where our existing explanations are incomplete. A first step in new research is to collect as many relevant facts as possible. To learn about the observations made by others, scientists read the primary literature and communicate with their colleagues. They also make their own observations, often aided by specialized equipment such as microscopes. Look for evidence of observations in the Introduction. How does previous work form the basis for the current study? What aspects of the research area are incompletely understood? What new observations led the scientists to undertake the work?

Some studies are mainly observational; for example their purpose might be to sequence a bacterial genome or to survey the species in a region of rainforest. In such cases, ask yourself whether the Introduction justifies the collection of new information. How do the new observations improve upon previous ones? Will the new data set be more complete or detailed? Is some new observational tool or technique available? Will the new data lead to new explanations or revision of current ones?

In some instances, specific research questions arise from observations. Suppose you observe that a bird is capable of unusually fast flight. What questions arise from this observation? One type of question asks *how* the bird achieves rapid flight. What anatomical, physiological, and biochemical properties contribute to its speed? Another type of question asks *why* the bird flies fast. Is the function to avoid predators, attack prey, migrate quickly, or some combination? New questions also arise when observations contradict expectations. Scientists recently found blood vessels in a *Tyrannosaurus rex* skeleton, countering the conventional wisdom that dinosaur soft tissue is not preserved and leading to a whole new set of questions about dinosaur evolution, anatomy, and physiology.

Choosing which questions to pursue is a crucial decision for scientists. Scientists may share a new research idea with other members of their research group. They will assess the scientific merits of the project. How important will the results be? Will a key research problem be resolved? They will consider how likely it is to succeed. Are there major technical hurdles to be overcome? Does the group have the time and money to pursue the project? Will it be possible to get funding? Finally, they will evaluate practical issues. Is another research group pursuing a similar project? Would the project help a student move towards completion of a degree? Will someone need to spend weekends taking care of animals or cell lines? In the end, both scientific and practical factors influence whether a project is pursued.

When scientists decide to tackle a research question, they carefully gather relevant observations and then propose an explanation. Scientific explanations take the form of **theories** and **hypotheses**. All scientific explanations are subject to revision, but theories are generally well-established explanations, while hypotheses are usually tentative explanations that have not been fully tested. Also, theories have broader implications than hypotheses, which tend to have a narrower focus.

The chief purpose of some papers is to synthesize observations and evidence into a new theory. If so, consider whether the authors justify the need for it. What biological area does the theory address? Are there existing theories that cover the same area? What are the inadequacies of existing theories? Also examine the strategy for developing a new theory. Why is this paper's approach better than that of previous workers? What experimental data or observations form the basis for the theory? What strategy or logic will be used to develop it? Will a mathematical model be constructed? Is there a plan to test the accuracy of the theory against the available evidence?

Single studies rarely attempt to test entire theories; more commonly studies test specific hypotheses. Hypotheses are tentative explanations. They are based on evidence accumulated through experiments and observations, and they are influenced by the prevailing theories. Good hypotheses lead to specific predictions that can either be contradicted or supported by experiments. When scientists repeatedly obtain experimental results inconsistent with the predictions of a hypothesis, then it must be discarded or substantially revised. When experimental results are consistent with predictions of a hypothesis, it gains support and can be tested further. Those that withstand rigorous and repeated testing become well accepted.

Hypotheses are often stated towards the end of the Introduction. They may be stated explicitly: "We hypothesized that drug X lowers blood pressure by dilating blood vessels." Sometimes they are stated without using the term hypothesis: "We tested whether drug X lowers blood pressure by dilating blood vessels." Or they might be stated as competing possibilities: "One possibility is that drug X lowers blood pressure by dilating blood vessels. However, it is also possible that it lowers blood pressure by causing increased urinary output."

Carefully consider the hypothesis. The Introduction should relate it to previous research. Have the authors made a convincing case for the importance of testing the hypothesis? Does it follow logically from prior research results? Also assess the relationship between the hypothesis and the theory. Is the hypothesis so central to the theory that its rejection would be a challenge to the theory itself? Or does the hypothesis deal with a peripheral aspect of the main theory? If the Introduction contains an outline of how the authors plan to test predictions of the hypothesis, think about the relationship between the experimental strategy and the hypothesis. Does the strategy constitute a rigorous test of the hypothesis? Is it possible to imagine a finding that would contradict predictions of the hypothesis?

SCIENCE AS A CYCLICAL PROCESS

Observation, explanation, and experiments form an interconnected cycle (Figure 2). Observations and experimental results make up the accumulated evidence that is synthesized into explanations, including theories and hypotheses. These explanations are tested by experiments and also motivate new observations. So the accumulated evidence guides the theories and hypotheses that scientists propose. In turn, theories and hypotheses guide the experiments scientists perform and the observations they make. As this cycle proceeds, evidence accumulates and theories become increasingly refined.

Reasoning in science is sometimes from general to specific and sometimes from specific to general. Development of theories moves from specific observations and results to general explanations with far-reaching implications. In contrast, testing of theories moves from the general to the specific as the reasoning proceeds from broad theoretical frameworks to more specific hypotheses and finally to even more specific predictions about experimental results.

You will occasionally read a paper that does not fit neatly into the framework we've discussed. Scientists draw on diverse tactics, and they may apply more than one strategy in a study. So, use this section as a guide towards understanding the Introduction, but don't be surprised to find creative approaches that defy this simple framework. And remember that consensus in science comes from the accumulated findings of many studies, so it is always wise to assess how each study relates to other research.

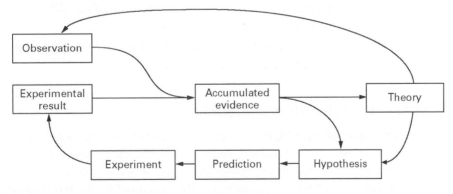

FIGURE 2. A simple model of scientific methodology.

SECTION 4 EXERCISES

Using a research article as an example, complete the following exercises:

1. Read the Introduction section of the paper. What is its main research area?

2. List the 10 most important terms used in the Introduction and give a definition for each term.

3. Identify two or three key previous studies that are described in the Introduction. Describe this previous work in your own words.

4. Are there areas of controversy in the research area? If so, what are they?

5. What new research question does the paper address? Why is this research question important? How does it extend previous work?

6. Does the paper test a hypothesis? If not, go to question 7. If so, restate in your own words the study's hypothesis. How does the hypothesis relate to the main theory? If the hypothesis is rejected, will the main theory be challenged? What predictions of the hypothesis are tested in the study? Describe a finding that would contradict the hypothesis and one that would support it.

7. Does the paper aim to develop a new theory or refine an existing one? If not, go to question 8. If so, is there an existing theory that addresses the research question? What are the shortcomings of the existing theory? What information, experimental or observational, is available to guide development of a new or refined theory? What approach does the paper take towards the refinement or development of theory?

8. Does the paper aim to collect a new set of observations? If not, go to question 9. If so, describe the new set of observations. How will they extend previous observations? Has a new technique or technology made collection of new observations possible? What new explanations could arise out of the observations?

9. Overall, does the Introduction section make a convincing case for the importance and value of the study?

Section 5: The Materials and Methods

The Materials and Methods section, sometimes simply called the Methods, tells how a study was performed. Authors describe the preliminary work, the experimental details, and the experimental design. When a study's methods are well documented, other scientists can repeat the experiments. We'll discuss in this section how to assess the effectiveness of a study's methods.

PRELIMINARY WORK AND APPROVALS

Scientists often conduct preliminary studies before they do the work reported in a paper. They may need to troubleshoot equipment or develop new methodologies. Preliminary experiments may be needed to optimize procedures. Patience, determination, and problem-solving skills are prerequisites for success during these early stages of a research project. If the authors discuss optimization of techniques or development of new procedures in the Materials and Methods, extensive work may have preceded the reported experiments.

Scientists need permission before starting some studies. If a study involves vertebrate animals or human subjects, scientists must apply to a **review board**. These boards are usually composed of other scientists, ethicists, and members of the general public. Review boards that evaluate human studies assess the balance between benefits and risks. In assessing benefits, the board considers whether the study will produce important findings. Are there general societal benefits? Will the subjects directly benefit, such as in a study that tests a disease treatment? Is the experiment well designed? Has the work already been done? In assessing risks, the board considers the chance of physical or psychological harm to the subjects. Will the researchers undertake every reasonable measure to minimize harm? Will subjects be completely informed of the study's procedures and risks and then asked to give their **informed consent**? Will subjects be unfairly coerced into participating, for example through unreasonable financial incentives or by the implication that participation is a condition of employment?

The review boards that assess animal studies are usually distinct from those that assess human studies. However, they consider many of the same factors. Animal review boards also ask whether the scientists have considered alternate models that would reduce the number of animals used. For example, could tissue culture replace some of the animal studies? If a study was approved by a review board, this is usually stated in the paper, often in the first few paragraphs of the Materials and Methods. Look for evidence of such review, especially if you have questions about the treatment of animals or human subjects. Other aspects of a study may also require approval, for example use of controlled drugs or radioactive materials. Again, look for evidence of such approvals in the Materials and Methods.

NUTS AND BOLTS: EXPERIMENTAL DETAILS

The Materials and Methods is often packed with technical terminology and methodological detail, sometimes making it difficult to read. How can you handle this dense information? One strategy is to evaluate the variables that were assessed, measured, manipulated, or controlled. In most studies, you will find three separate types of variables: dependent variables, independent variables, and controlled variables.

Dependent Variables

Dependent variables change in response to other variables. The purpose of many experiments is to characterize dependent variables and how they change under different conditions. For example, in a study of volume regulation in fish cells, cell volume is the dependent variable. Experiments might examine how it changes in response to other variables, such as the osmolarity of the surrounding fluid. Take care to identify the dependent variables, because these are often the study's focus.

Measurement of dependent variables may require sophisticated techniques. Be sure that you understand the key techniques used in a paper. You can't evaluate the results if you don't know how they were obtained. Some techniques involve a number of steps. Suppose for example a study examines changes in gene expression during cell volume regulation. A first step would be to isolate RNA, a process which includes several centrifugation and incubation steps. Relying only on the text of the Materials and Methods, you might find it difficult to see how each part of a multi-step procedure fits together. To better understand a technique, try to draw a flowchart that outlines its steps. Figure 3 shows how such a flowchart might look for a standard RNA isolation procedure. When you depict a technique this way, the methodological details as well as the overall strategy become apparent. Basic procedures may not be described in the Materials and Methods because the authors assume the audience is familiar with them. In such cases, you may need to consult secondary sources to learn more.

Independent Variables

Independent variables potentially influence the dependent variable. In some cases they are directly manipulated by the experimenter. Other times they are measured but not manipulated. Let's consider again a study of cell volume regulation. Osmolarity of the fluid outside cells is an independent variable because it is predicted to affect cell volume. Osmolarity can be manipulated, for example, by placing cells into solutions of different solute concentration. Suppose, however, we were interested in whether or not cell volume regulation is different in the cells of marine versus freshwater fish. In this case, the independent variable is the habitat of the fish. Although we may be unable to change this variable, we could still examine its effect by studying fish collected from different habitats. Some studies test multiple independent variables; be sure to identify all of them.

Controlled Variables

A difficulty in many studies is that numerous factors other than those under investigation could affect the outcome. For this reason, scientists seeking to measure the influence of independent variables strive to control other variables; these are called **controlled variables**. For example, studying the effect of osmolarity on cell volume requires that other factors, such as temperature and pH, are held constant. Variables can sometimes be controlled through experimental design. Imagine that

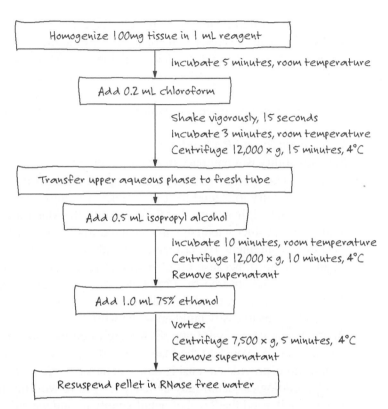

FIGURE 3. Flowchart of a standard RNA isolation procedure, modified from instructions for Trizol reagent, Invitrogen.

a difference was detected between cell volume regulation of a marine compared to a freshwater fish. Can this difference be attributed to the different habitat or is it due to some other difference between the two species? Studying a single species, such as salmon, that migrates between marine and freshwater environments would make species a controlled variable. Scientists seek to control as many variables as they can, because doing so enables them to draw stronger conclusions. Look closely at the Materials and Methods to see which variables the investigators have controlled. Consider whether there are other variables that could have influenced the results but were not controlled.

Controls

Many experimental designs use **control groups**. Don't confuse these with controlled variables; they're not exactly the same. Control groups usually receive a treatment where the independent variable is unchanged from the normal or ordinary value. They can therefore serve as a comparison for the **experimental groups** that receive a different treatment. Ideally, the control and experimental groups are treated identically except for the difference in the independent variable. If so, all variables except the independent variable are controlled and any difference between groups can be attributed to the independent variable. Unfortunately, this ideal situation is rarely found in actual experiments, and you should carefully consider control groups to see how well they mirror the experimental groups.

Reproducibility and Repeatability

Reproducibility is a hallmark of scientific progress. Scientists repeat the work of others because verifying a study's findings strengthens its conclusions. Scientists also regularly try to extend their colleagues' findings, and a first step is to repeat and verify the original work. For these reasons, the Materials and Methods must be written so that other scientists can repeat the studies. Extensive detail is required to meet this requirement. Experimental conditions and technical detail must be thoroughly described, because even small differences may alter outcomes. Standard methods may be covered by a reference to a previous paper, but any deviations from them must be noted. When authors fully describe their methods, they enable their colleagues to easily build upon the work. This can save other scientists huge efforts, because successful approaches are often the outcome of time-consuming optimization. As you read the Materials and Methods, ask yourself whether sufficient detail is provided to enable other scientists to repeat the study.

Methodological detail may be useful if you're working in a research lab, but do you really need that level of detail simply to understand a paper? Perhaps not. But you will still sometimes need to consider procedural details. Consider two papers that come to contradictory conclusions. Examining how each study was performed might uncover differences that explain the discrepancies. You may also need to refer to the Materials and Methods as you assess the Results. Suppose a paper reports that "women who ate a high-carbohydrate diet ran 15% longer than those who ate a normal diet." You would want to know many methodological details before interpreting this result. How many women were tested? What were their ages? Were they trained athletes? How were they selected to participate in the study? What was the composition of the diet? How fast did the women run? The Materials and Methods section should supply these details. One way to use the Materials and Methods is to consult it for details that are necessary to interpret the Results.

THE BIG PICTURE: EXPERIMENTAL DESIGN

Convincing findings arise when a study's design is well matched to its purpose. As we'll see below, different purposes call for different experimental approaches. In reading the Materials and Methods, one of your main tasks is to assess the experimental design's effectiveness. Correlative studies are most appropriate for describing the **patterns** in nature. Causative studies are most effective at establishing the **causal factors** that explain those patterns.

Correlative Studies: Patterns and Connections

In **correlative studies**, scientists do not manipulate independent variables, but instead exploit preexisting variation in them. For this reason, they are often called **retrospective studies**. They are also sometimes referred to as **cross-sectional studies** or **observational studies**. Because correlative studies can investigate multiple independent variables simultaneously without artificial manipulation of the conditions, they are an excellent means for identifying connections between variables.

Let's look at an example. Suppose that scientists suspect that an insecticide, used on vegetables, causes cancer in humans. A correlative approach would be to identify people who have been exposed to the insecticide and compare their cancer rates to a control group. Cancer rate is the dependent variable in this study, while insecticide exposure is the independent variable. If cancer rates are the same in both populations, then the insecticide probably does not cause cancer. If cancer rates are higher in the exposed population, a correlation has been established between cancer rate and insecticide exposure.

Correlative Studies: Challenges and Limitations

A challenge in correlative studies is that many independent variables may be correlated with the dependent variable. For example, family history, diet, exercise, and smoking all might affect cancer rates. Interpretation is complicated if multiple factors are different between the control and experimental groups. Thus, scientists attempt to match the groups, so that variables other than those being studied are the same. If scientists can't match the groups, they measure variables known to influence the dependent variable and use statistical techniques to account for the differences. In assessing a correlative study, consider whether the investigators have accounted for all the variables that could influence the outcome.

Correlative studies cannot demonstrate causation; they can only suggest it. One issue is that cause and effect can be confused. Consider a study that demonstrates a correlation between exercise and healthy heart function. One possibility is that exercise causes improvements in heart function. But it is also possible that people with healthy hearts are more likely to exercise; in this case, a healthy heart contributes to high levels of physical activity rather than the reverse. Another issue is that independent variables may be correlated with changes in a dependent variable even if they're not causative. In the insecticide example, consider the possibility that other cancer risk factors are present in the part of the country where the insecticide was used. If so, the insecticide could be associated with cancer risk only because it is correlated with a causative factor, not because it is a cause itself. Correlative and causative studies are often used together. Correlative studies can point out possible causes, which can then be investigated in causative studies.

Causative Studies: Between-Groups Design

In **causative studies,** scientists manipulate an independent variable and measure the effect on a dependent variable. In a **between-groups design**, experimental groups receive a treatment while a *separate* control group does not. For example, in Figure 4A, the experimental group exercises and the control group does not. The effect of the experimental treatment can be determined by comparing post-treatment measurements in the control and experimental groups. Random assignment of subjects into groups minimizes the chance that they differ in factors other than the manipulated independent variable. Measurements on experimental and control groups can be made at the same time under identical conditions, ensuring that variables other than those under study are held constant. However, because separate individuals are assigned to control and experimental groups, between groups designs do not control for interindividual variability. When such variability is high, the groups are more likely to differ from each other prior to the treatment, and it can be difficult to differentiate between preexisting differences and those due to the treatment.

Causative Studies: Repeated-Measures Design

Repeated-measures designs control for differences among individuals by studying the *same* subjects after exposure to different treatments. A simple repeated measures design is to make measurements **before and after** a treatment (Figure 4B). The treatment effect can then be expressed as the *difference* between the before and after measurements in the same individual, controlling for interindividual variability. This makes it easier to detect differences, allowing smaller sample sizes to be used. Note that in the before-and-after design, time-related variables are not controlled since the treatment effect is always measured after the control. This can be a problem when the variables are influenced by the time of day or when making an initial measurement affects later measurements.

A **cross-over** repeated-measures design addresses these issues. Cross-over designs are similar to between-groups designs, except that the same individuals undergo both the control and experimental treatments on separate occasions. Treatments can be matched so that only the studied variables differ. To control for effects due to the order of treatment, subjects can be randomly assigned to receive the experimental or control treatment first. Figure 4A would depict a cross-over design if the same subjects performed the control and exercise treatments on separate days.

Model Systems

Causative studies often employ **model systems.** Models are alternative experimental systems that save time and money and allow experiments that would otherwise be impossible. A laboratory animal might be used instead of humans, or tissue culture might be used in place of a living animal. However, care must be taken in extrapolating from the model. Scientists choose model systems with great care, balancing their

A. Between groups design

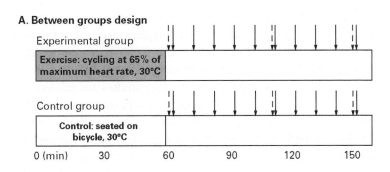

B. Before and after repeated measures design

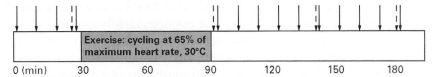

FIGURE 4. Timelines of two different experimental designs. Solid arrows represent blood pressure measurements; dotted arrows represent blood samples. Panel A shows a between-groups design if separate subjects form the control and experimental groups. The same timelines would depict a cross-over repeated-measures design if the same subjects received control and experimental treatments on different occasions. Panel B shows a before-and-after design.

advantages and disadvantages. If a study uses a model system, you should consider whether it makes the experiments easier to perform and also assess how well it represents the system it replaces.

Let's return to the example of insecticide and cancer risk. Conducting a causative study of insecticide use and cancer risk in humans may be impossible; exposing humans to a suspected carcinogen is clearly unethical. Instead, suppose that carcinogenicity could be assessed by applying the insecticide to cultured cells and assessing their production of a cancer marker. In this case, you would want to assess whether findings in these tissue culture experiments are applicable to insecticide exposure in humans. For example, is it possible that the insecticide is broken down when ingested by a human but is not when applied to cultured cells?

Evaluating Experimental Design
The Materials and Methods may not explicitly state what type of study was conducted or which groups are the controls. Sometimes, you can make sense out of the experimental design by diagramming it. Draw a timeline for each experimental group and indicate treatments and measurements on it (Figure 4). The study design and control groups should become obvious. Combine this with technique flowcharts (Figure 3), and you'll have a complete visual representation of the study.

Correlative studies may be the best approach when investigators seek to study the interactions of multiple independent variables, when it is important to study a system without manipulating it, or when the purpose is to identify possible causative factors. Also, correlative studies may be the only option if manipulating variables is impossible or unethical. Causative studies are needed to establish causation. If interindividual variability is large, a repeated-measures approach is usually best. For example, if dishes of cells varied considerably in their control level of cancer marker, sampling them before and after exposure to the insecticide would be a good approach. However, repeated-measures designs are not possible in circumstances when only one measurement is possible, for example if measurement of cancer marker damaged the cells. Repeated-measures designs also don't work well if the subjects are changing rapidly over time because it is then possible that they've changed between treatments. For example, rapidly growing cells might display different properties over time, making comparisons on separate occasions difficult. Try to determine why a particular study design was used. Consider the study's purpose as well as the limitations and constraints of the experimental system.

Critically consider the controls in causative studies. What would be an appropriate control treatment in the experiment of insecticide exposure in tissue culture cells? To decide we must know how the experimental treatment was performed. Exposing cells to insecticide might include preparing a solution of insecticide in solvent, pipetting the insecticide solution onto the cells, and then mixing by gentle shaking. A good control would match everything except the insecticide exposure, for example by pipetting a solution containing the same solvent without insecticide onto the cells. Check to see whether there are differences between the experimental and control groups other than the studied variables.

SECTION 5 EXERCISES

Using a research article as an example, complete the following exercises:

1. Was preliminary work done before the reported experiments were performed? How does the preliminary work relate to the reported experiments?

2. Did the authors obtain approvals from animal or human review boards or other regulatory agencies? Do you have questions or concerns about treatment of human or animal subjects?

3. List the variables studied. Differentiate between independent, dependent, and controlled variables.

4. How do the authors measure the dependent variables? Were the independent variables manipulated by the investigators? How were other variables controlled? Did the investigators fail to control any important variables?

5. Choose a key technique and draw a flowchart to depict it.

6. Do the Materials and Methods provide enough detail for another scientist to repeat the work?

7. Draw a timeline depicting the experimental design. Indicate the timing of measurements and treatments.

8. Was a model system used in the study? If so, what experimental advantages does it have over the system it replaces? How well does it mimic the system it replaces?

9. Describe the overall study design. Classify it as correlative or causative. If it is causative, is it a repeated measures or a between-groups design? Does the design fit well with the study's main purpose?

Section 6: The Results

The Results, together with the Materials and Methods, are the core of a study. New observations, data, and findings are presented in the Results. In this section, we discuss strategies to help you assess the primary data.

Scientists collect data in many forms, including numerical output from instruments and visual information such as photographs and micrographs. Unprocessed data must be recorded in a timely, accurate, and lasting form. Human memory is not always trustworthy, and thus the most reliable records are made immediately. Scientists keep a lab or field **notebook** where they record their methods and data. You can view the Results section as a translation of unprocessed data from the notebook into a succinct and easily understood form. Rarely does this mean sharing all the data in exactly the form they were collected. Data are analyzed, sorted, and synthesized before they are presented. The notebook is also important as physical proof that the work was actually performed. In the rare case of an accusation of research fraud, it becomes an important piece of evidence.

DEALING WITH VARIABILITY: STATISTICS

In this section, we discuss basic statistical principles needed to assess biological studies. When reading articles, you may need to consult a statistical manual for more information about specific statistical tests (see Resources, p. 43).

Variability is ubiquitous in measurements of biological systems. **Technical variability**, or **measurement error**, arises any time a measurement is made. No technique is perfect, and variability is introduced when measurements deviate unpredictably from actual values. Scientists always seek to minimize technical variability because it complicates data interpretation, but it can never be completely eliminated. **Biological variability** refers to real differences between individuals. Because biological variability represents actual differences, there's no way to completely eliminate it. In fact, scientists sometimes want to study it because it plays a key role in processes such as natural selection. Although biological variability can't be eliminated, careful experimental design can minimize its influence. For example, when studying animals, matching factors such as age, weight, and sex can reduce biological variability.

Due to variability, biologists need to make multiple measurements to fully characterize a system. These measurements constitute a **data set**. Sometimes an entire data set is shown in a paper. If the heart rate of six cyclists has been measured, they might all be presented in a table. More commonly, data sets are summarized. Sometimes, data sets are presented graphically, as **histograms** or **frequency distributions**. Some distributions are **symmetrical** (Figure 5A). **Normal distributions** are a specific type of symmetrical distribution that form a characteristic bell-shaped curve. Many statistical tests assume a normal distribution, a condition not always met in biological data sets. Nonsymmetrical distributions can have various shapes. For example, they may be skewed in one direction (Figure 5B). Distributions may also be **unimodal**, having a single peak (Figures 5A and 5B), or **multimodal**, having more than one peak. In evaluating data, try to get a sense of the way the values are distributed. Unfortunately, this can be difficult to assess in many papers, because the needed data are often not provided.

Descriptive Statistics: Central Tendency and Variability
Data sets can also be summarized by describing two features: **central tendency** and **variability**. Central tendency describes the typical or representative value. **Means**, the arithmetic average of the points in a data set, are the most common way of representing central tendency. They work particularly well with symmetrical data sets. **Medians** are the middle value when the points are arranged from highest to lowest; **modes** are the most common value in a data set. Medians and modes can be useful with nonsymmetrical data sets, such as those that are skewed in one direction. In these data sets, the most extreme outlying values can have a disproportionate effect on the mean, pulling it away from the central tendency. Most papers in biology summarize data using means. Be aware that means are only a partial representation of a data set. The shape of the distribution and the amount of variability are also important.

Two common measures of variability are the **range**, the minimum and maximum values, and the **standard deviation,** which represents an adjusted average distance between individual data points and the mean. A large standard deviation indicates high variability, meaning the data are more spread out compared to a data set with a small standard deviation. For a normal distribution, about 68% of the values are within one standard deviation of the mean, while about 95% of the values are within two standard deviations. Values are often reported as a measure of the central tendency ± a measure of variability; for instance mean ± standard deviation. In assessing results, consider the magnitude of the variability in the context

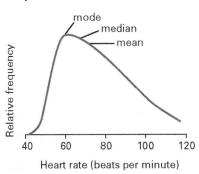

FIGURE 5. Frequency distributions. Relative frequency is the proportion of all the values in a data set that fall into a particular range. Panel A shows a symmetrical distribution. Panel B shows an asymmetrical distribution with a positive skew.

of the central tendency. For example, a standard deviation of 1 second may be irrelevant when comparing a mean difference of several hours, but crucial when evaluating a mean difference of a few seconds.

Inferential Statistics: From a Sample to a Population
It's often desirable to generalize from a studied group, the **sample**, to a broader group, the **population**. For example, scientists are probably not just interested in how *one particular* flask of *Escherichia coli* responds to altered glucose concentration; more likely they hope their findings will apply to other cultures of *E. coli* as well. **Inferential statistics** enable scientists to generalize from a specific sample to a wider population. Uncertainty always exists in this process, and inferential statistical approaches therefore lead to statements of *probabilities* rather than certainties.

   If you want to know the mean heart rate of women at your college, one approach would be to study every woman on campus. If this was impractical, an alternate approach would be to study a properly selected sample and then generalize to the broader population. If heart rate was measured in 10 women, the mean of this sample would be an estimate of the population's mean. Repeating the measurement with ten different women would probably return a slightly different mean. The results in most papers reflect a similar scenario; some individuals are sampled as representatives of a larger population. A crucial step in assessing such results is determining how accurately the sample reflects the population. First determine what population the scientists are interested in, and then assess how well the sample represents this population. For example, a sample of varsity athletes would probably not be representative of the heart rates of the general campus population, while a random sample of enrolled students would be more representative.

The reliability of an estimate is also affected by the variability and the **sample size**, the number of individuals measured. When variability is high, a large sample size is required to get a good estimate of the mean. Conversely, a smaller sample size is adequate when variability is lower. **Standard error** accounts for both sample size and variability and is commonly used to represent uncertainty in an estimate of the mean. As standard error grows smaller, the likelihood grows that the sample mean is an accurate estimate of the population mean. Try to consider both sample size and variability when evaluating the reliability of an estimate. If there is uncertainty, is it because of high variability or a small sample size?

Statistical Tests: Null and Alternative Hypotheses
Suppose you sampled 20 men and 20 women on your campus and found a difference in heart rates between sexes. One possibility is that a real difference exists between men and women on campus. Another is that the measured difference occurred by chance as a result of sampling only part of variable populations; for example, the sample of women might have inadvertently included several women with higher than average heart rates. Inferential statistics can help distinguish between these possibilities by determining the *probability* that a difference in sample means is due to a true difference in the population.

Many statistical tests differentiate between a **null hypothesis** and an **alternative hypothesis**. These hypotheses are constructed so that only one can be true. For example, if the null hypothesis states there is no difference between treatments, the alternative hypothesis states that there is a difference. The alternative hypothesis is the one that requires strong support in order to be accepted. To establish a rigorous test of the alternative hypothesis, statistical tests begin with the assumption that the null hypothesis is true. The alternative hypothesis gains support only when the null hypothesis is rejected. The bottom line is that the alternative hypothesis is rigorously tested while the null hypothesis is not.

Statistical hypotheses are only valid when they are developed before data are collected or examined. It's always possible to develop a hypothesis that fits a particular data set after it is collected, but when this is done the hypothesis has not been rigorously tested. Assessing whether a hypothesis was developed before or after conducting the experiments can be difficult. One clue is whether it was mentioned in an earlier paper.

The alternative hypothesis usually states the interesting result: that there is an effect, difference, or correlation. Consequently, the null hypothesis usually states that there is no effect, no difference, or no correlation. Since the alternative hypothesis is rigorously tested, an interesting result is only accepted when there is strong support for it. Suppose we are interested in determining whether there is a difference between the heart rates of men and women. The null hypothesis would be that there is no difference, and the alternative hypothesis would be that there is a difference. If the null hypothesis is rejected, a difference between men and women would be strongly supported. Papers often state only the alternative hypothesis, but you need to understand both the null and alternative hypotheses to assess statistical tests. If the statistical hypotheses are not explicitly stated, try to determine them yourself.

Positive Results

Rejection of the null hypothesis is a **positive result**, because the alternative hypothesis is strongly supported. However, statistical tests do not reject null hypotheses with absolute certainty. There always remains the possibility that a null hypothesis has been mistakenly rejected. Rejecting a true null hypothesis is a **Type I error**, also called a **false-positive error**. Biologists are willing to accept only a low possibility of making such errors, because they can lead to the erroneous acceptance of the interesting result described by the alternative hypothesis.

Inferential statistics assess the probability that a false-positive error will be committed. Many statistical tests return a number called a **p-value**. Findings are labeled as **statistically significant** when the p-value is less than a preestablished **significance level**. A typical significance level in biology studies is 0.05; this means that the null hypothesis will be rejected if there is less than a 5% chance of doing so mistakenly. When considering statistically significant results, always assess the probability that an error has been made. If p-values are reported, they directly indicate the likelihood of making a false-positive error. Very low p-values indicate that the null hypothesis can be rejected with high certainty. In other words, as p-values decrease, the chance of making a false-positive error also decreases.

Also consider the significance levels used in a study. A low significance level ensures a small chance of committing a false-positive error. For example, a low significance level might be warranted when testing a risky drug treatment in an experimental animal, because a very high likelihood of effectiveness might be needed prior to human testing. On the other hand, less certainty might be acceptable in preliminary studies with high variability or lower sample sizes.

Because each study has some probability of making a false-positive error, such errors inevitably slip into the published literature. Thus, some of the statistically significant findings reported in papers are false-positive results. Reproducing findings is an antidote to this problem. Statistical significance is unlikely to arise erroneously in several experiments, especially if they are done with different conditions or methodologies.

Negative Results

Negative results arise when the null hypothesis is not rejected; in such cases the alternative hypothesis is not supported. As with positive results, it is always possible that a negative result is mistaken. Failing to reject a false null hypothesis leads to a **Type II error**, also called a **false-negative error**. In this case, an alternative hypothesis that should have been accepted is not. In contrast to a false-positive error, the probability that a false-negative error has occurred usually cannot be determined. Therefore, when the null hypothesis is not rejected, it is not appropriate to accept it or to reject the alternative hypothesis. In fact, the evidence may even favor the alternative hypothesis. For example, in the case of a p-value of 0.06 in a study where the threshold is set at 0.05, the null hypothesis would not be rejected. Also, a failure to reject the null hypothesis might occur because the experimental design was flawed or the sample size was too small. Consider the possibility of false-negative results when differences are found to be not statistically significant.

Negative results can be hard to publish. Convincingly demonstrating no difference between treatments or no connection between variables is difficult. Also, findings of no effect generally don't generate the same excitement as positive results. But there are good reasons to publish well-designed studies that produce negative results. One reason is that it saves other scientists from repeating the same study; another is that finding no effect might have important biological implications. Ruling out one cause might strengthen the case for another.

Biological Relevance

A finding of statistical significance does not necessarily imply that a result is biologically meaningful. Suppose that two species have a statistically significant difference in mean body temperature: 37.1°C for one species and 37.2°C for the other. Does a 0.1°C difference in body temperature have any important biological implications? Conversely, further study of a biologically interesting difference, such as a 3°C difference in body temperature between species, might be worthwhile even without proof of statistical significance. Remember that false-negative results are possible. Statistical significance might arise with experimental refinements or increased sample sizes. As you assess the results, look carefully at the magnitude of differences. Are there large differences that fail to reach statistical significance? Do statistically significant differences vary enough for there to be biological consequences?

VISUALIZING RESULTS: TABLES AND GRAPHS

Results can be presented as pictures, graphs, tables, or as statements in the text. Consider how data are presented; this gives you a sense of which results the authors wish to emphasize. Tables can concisely present large data sets, but it's tough to emphasize particular findings using a table. Presenting data in graphs takes a bit more space than in tables, but graphs are more effective in illustrating differences. The most important findings are ordinarily presented in graphs and tables, because these attract the most attention from readers. Key findings are also emphasized in the Results text. We'll focus below on graphs; you can apply a similar approach to reading tables.

Graphs

Take your time assessing graphs, because they consolidate large amounts of information. A graph's purpose can be discerned by determining the dependent and independent variables and reading the **figure legend**. Independent variables are ordinarily shown on the **x-axis**, and dependent variables are plotted on the **y-axis**.

Imagine that investigators have identified a new species of bacteria from a hot spring and have tested its growth rate at two different temperatures (Figure 6). In this graph, specific growth rate is the dependent variable, temperature is the independent variable, and comparison of growth rates at two temperatures is the purpose. Consider how the purpose of the experiment plotted on a graph relates to the study's overall purpose. For example, why is the comparison of growth rates shown in Figure 6 important? You may need to return to the Introduction for a reminder of the study's goals.

Examine the graph's **units** and **axis scale**. First, determine the unit of measurement, which is often given in parentheses after the **axis label**. Notice that specific growth rate in Figure 6 is expressed as h^{-1}. What does this unit mean? How was growth rate measured? What experimental conditions were used? You will find answers to some questions in the figure legend, but you may also need to check the Materials and Methods. Also assess what range of values was measured by looking at the y-axis scale. In Figure 6, specific growth rates range from about 0.02 h^{-1} to 0.04 h^{-1}. Are these high or low growth rates? You may need to compare values to those from other figures or even other papers in order to get a sense of what the reported numbers mean.

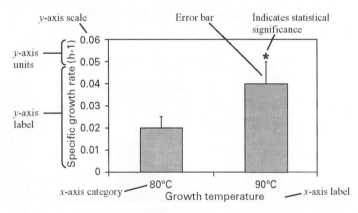

FIGURE 6. Growth rate of the fictional hot springs bacteria *Bacterium warmus* measured at 80°C and 90°C. Bars represent means ± standard error. *n* = 12 for each species. Significant differences between treatments (p < 0.05, Student's *t*-test) are indicated by a (*).

Once you understand the purpose and methods behind a graph, try to discern the major patterns in the data. What are the major trends? Are the differences between treatments large or small? Consider measures of variability, which are often given as **error bars** projecting above and/or below data points. How variable are the data? Is the variability consistent across treatments? Do any data points fail to follow the overall trend? Read the figure legend closely; it often gives important information such as the sample size (often abbreviated as n), the type of variability shown in the error bars, and how the data were statistically evaluated. In Figure 6, the sample size was 12, the error bars depict standard error, statistical differences were evaluated with a Student's t-test, and the significance level was set at 0.05. Sometimes statistically significant findings are indicated by marks on the graph. Read the figure legend to learn how these marks are used. In Figure 6, the * symbol above the 90°C bar is used to show that the difference in growth rates at the two temperatures is statistically significant.

Evaluating Data

As you evaluate the results, consider how they've been presented. Let's suppose growth rates of a second hot springs bacterial species were also measured. The graph in Figure 7 emphasizes the difference between the species. The effect of temperature on *Bacterium hottus* is also apparent. But notice that the effect of temperature on *Bacterium warmus*, which was evident in Figure 6, is now difficult to discern in Figure 7 because the y-axis scale has been changed to accommodate the higher growth rates of *Bacterium hottus*. The bottom line is that a graph can emphasize or camouflage different aspects of the data. When reading a graph, ask yourself how the data are presented. Are certain differences highlighted? Are others obscured? Does the presentation reflect the authors' opinions about the data? Do you see interesting patterns that are downplayed in the authors' presentation? Does the figure legend help clarify the graph?

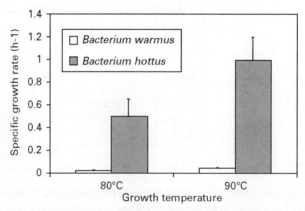

FIGURE 7. Growth rates of two fictional hot springs bacteria measured at 80°C and 90°C. Bars represent means $\pm$ standard error. $n = 12$ for each treatment.

Try to evaluate the numerical data without being influenced by how they are presented. One method is to examine the graphs and tables for major trends before reading the Results text. Another strategy is to read the Results section text first, picking out the main statements the authors make about their data. Then, check to see if the numerical data in the tables or graphs support these statements. The key is to develop your understanding of the study's findings and then compare it to what the authors say. If there's a discrepancy, you might have identified a shortcoming in the authors' analysis. On the other hand, you might also have misunderstood the authors, so check your analysis carefully.

Why go through all the trouble to evaluate the primary data yourself? Why not simply read the Results and accept what the authors say? The answer is that only after evaluating the primary data can you critically assess the study and its conclusions. Just as you can't intelligently review a movie or a book unless you've read it yourself, you can't critically assess a scientific study without personally evaluating its core component: the data.

SECTION 6 EXERCISES

Using a research article as an example, complete the following exercises:

1. How are the data presented in the paper? In pictures, graphs, tables, or text?

2. Choose a key table or graph and use it to answer the following questions. What are the main trends in the data? Are the differences between treatments large or small? How is variability depicted? How variable are the data? How are the data distributed? How is central tendency presented?

3. Summarize each table or graph in the paper in a few sentences. Compare your description to that of the authors.

4. What population is the study interested in examining? Does the study sample only part of the population? Is the sample representative of the entire population?

5. How well do the measurements of the sample estimate the properties of the population? If possible, use standard errors or another statistical measure to support your answer.

6. Choose a key experiment that was analyzed with statistics. State the null and alternative hypotheses. Describe the outcome of the statistical test. If a statistical significance was found, what is the likelihood that a false positive occurred? If no significance was found, can you assess the possibility of a false negative?

7. Are the findings biologically relevant? Are there any findings that didn't reach statistical significance but might be worth further study?

Section 7: The Discussion

The Discussion is an opportunity for authors to explain what their findings mean, to illuminate the key conclusions, and to address potential criticisms. Information is drawn from many different sources, and opinion may mingle with objective fact. Your task is to sort through this information and come to your own conclusions about the study. Be prepared to consult other writings as you read the Discussion. Because it synthesizes information from the rest of the study, you may need to refer back to the Introduction, Materials and Methods, and Results. Since it connects the study to previous work, you may need to consult other research articles.

INTERPRETATION: FINDING MEANING

The Materials and Methods and Results mostly report factual information. While superficial interpretation may be found in the Results, in-depth interpretation is usually confined to the Discussion. **Interpretation** differs from simple reporting of experimental results because it involves describing the *meaning* of the data. Evaluating how well the study fulfills its purpose is a form of interpretation. Authors may describe whether a useful data set was collected, whether the findings support or contradict the hypothesis, or whether a new theory was developed. Another type of interpretation is the synthesis of a study's different findings. Authors may state whether all the study's findings are consistent, whether any results contradict the others, or whether more reliable conclusions can be made when a set of experiments is considered as a whole. Interpretation also includes describing the strengths and weaknesses of a study. Authors may address the shortcomings of their methodologies and the limitations of their conclusions.

Critically reading a paper requires that you completely understand the authors' interpretations, and then compare them to your own. This may seem like a daunting task. Is it even possible to question the interpretations of the authors? Aren't they in a much better position to make interpretations than you? Although you will generally have no basis to question the actual data, you will be able to evaluate *interpretations* of the data. If the authors have presented their methods and results clearly, you have access to all the information that is necessary for developing interpretations. Furthermore, it is possible to arrive at different interpretations than the authors. For example, you may approach the paper from a different perspective, you may have new information that has become available since the authors wrote the article, or you may be more objective in assessing the experimental design and results. A strength of the scientific process is that scientists vigorously challenge each other's interpretations. As you begin developing your own interpretations and assessing those of others, you become an actual participant in the scientific process.

How can you develop your own interpretations? Here's where the hard work you've done assessing the Introduction, Materials and Methods, and Results pays off. Before you read the Discussion, review the other sections with the aim of developing your interpretations. Write these down. Then read the Discussion to see how the authors interpret their findings. You may come across interpretations that didn't occur to you. In these cases, consider whether you agree with the authors, referring if necessary back to the other sections. Developing and assessing interpretations takes practice. You won't be an expert on your first try, but you'll improve with time.

CONNECTIONS: RELATIONSHIP TO OTHER WORK

The Discussion compares the study to previous research, placing the work into the context of a broader research field. This is a key activity, since consensus in science usually emerges from many studies considered together. Authors may discuss how their findings contradict other studies, how previous work supports their conclusions, and how their work extends the knowledge within a field.

Reading the Discussion can be complicated because authors interpret not only their work, but also others' work. Keep track of this as you read by looking for cues that indicate what is being discussed. Previous work is usually identified with a citation: "Physical activity level has been found to be correlated with blood pressure (Smith, *et al.*, 2003)." Or, "Smith and colleagues (2003) found that physical activity level was correlated with blood pressure." Also consider whether a previous study is simply being summarized, as is the case in the examples above, or whether the authors are interpreting or criticizing the study. "Because Smith and colleagues (2003) studied only 10 college-aged males, care must be taken in applying their results to other populations" is an interpretation. Be careful about accepting criticisms of previous work. In fairness to Smith and colleagues, you should check their paper before judging it.

Previous work is often used to support a study's conclusions. The strongest claims can be made when studies by different investigators are consistent. Claims are particularly strengthened when studies using different approaches come to the same conclusion, because it is unlikely that several approaches are flawed. Assess how the study relates to other work. Are the study's conclusions consistent with prior work? Does it use a different approach than previous work? Does the study come to a new conclusion, or is it another piece of evidence supporting a well-studied theory? Does it strengthen a previously shaky conclusion? Authors also describe whether their study conflicts with prior work. Assess such contradictions carefully. Were the previous studies somehow flawed? What differences in methodology may have led to the contradictory results? Why might the new study be more reliable? Here again you may need to consult some of the previous studies to get their perspective.

EXPLANATIONS AND IMPLICATIONS

Scientists have the opportunity to explain their results in the Discussion. For example, when a study demonstrates a cause-and-effect relationship, scientists seek to explain the **mechanism** that connects cause and effect. Suppose a study finds that exercise lowers blood pressure. This conclusion will be strengthened if we have a plausible explanation of how exercise causes a reduction in blood pressure. Sometimes the study itself gives possible clues to the mechanism. Other times mechanisms can be proposed based on previous studies. Look for mechanistic explanations in the Discussion. Do the authors give convincing explanations for their findings? Do their data suggest mechanisms? Are other mechanisms possible?

Some studies are specifically aimed at identifying mechanisms. In such cases you should assess how convincingly the mechanism has been established. Even when a mechanism is clearly demonstrated, a new set of mechanistic questions often arises. As studies accumulate, explanations become more detailed and accurate. The Discussion should give you a sense of this process. How sophisticated are the current mechanistic explanations? Are they well refined or are they general and approximate? Is there more than one competing explanation?

Authors usually describe the significance of their work toward the end of the Discussion. Studies can have scientific significance, practical applications, or both. A study might suggest a new set of research questions, put forward a new theory, or resolve a long-standing controversy. It might support the effectiveness of a new drug, improve a manufacturing process, or aid in the development of a new technology. Examine carefully any claims of significance in the Discussion. Are they justified by the information presented in the paper? Are the implications direct and immediate, or are they tentative and speculative? Sometimes authors describe future work in the Discussion. This can be a clue to the significance of a study. Important work usually leads to exciting new questions to explore.

SECTION 7 EXERCISES

Using a research article as an example, complete the following exercises:

1. How do the authors interpret their findings? Which results do they consider to be most important? Do they claim to support or contradict a hypothesis? Do they synthesize their individual findings into a coherent story?

2. Do the authors address weaknesses of their methods or findings? Do they address possible criticisms?

3. How does your interpretation of the study compare to the authors'?

4. Describe how the study connects to previous research in the field. Is it supported by prior work? Does it contradict any previous conclusions?

5. Do the authors propose an explanation for their findings? Is a plausible mechanism proposed? Is there any evidence to support the proposed mechanism?

6. What are the implications of the study on the research field? Does it suggest new work that needs to be done? What are the next steps that need to be accomplished?

7. What is your overall opinion of the study?

Section 8: Putting it All Together

We've covered lots of specific aspects about reading research articles. Let's think about the themes that have emerged. Here are 10 tips to guide your reading of the primary literature:

1. Focus on methods and results. Try not to be influenced by the way the study is presented, but rather focus your analysis on the experimental design, techniques, and data.

2. Be a skeptic. Ask yourself how strongly the authors' interpretations and conclusions are supported by the evidence.

3. Be fair. Scientific research is difficult, and scientists operate under many constraints. Don't expect studies to be perfect.

4. Read nonlinearly. Exploit the format of research articles to quickly access the information you need. Don't feel compelled to read every line start to finish. Skim the paper to understand its overall approach. Refer to previous sections as necessary.

5. Consider the big picture. Assess where the study fits into the cycle of science, and how it relates to previous research.

6. Consult other sources. Writers of research articles assume their audience has basic knowledge of the area. Consult secondary sources to get the needed background.

7. Take your time. Research articles condense entire studies into a few printed pages. It probably took the authors years to conceive, perform, and publish their work. Be patient and persistent when reading articles.

8. Accept uncertainty. Research articles deal with emerging knowledge and controversial issues. Don't expect to find absolute answers to every question. Each paper is a step in an ongoing process.

9. Expect to be challenged. If you're not an expert in an area, there might be aspects of a paper you can't understand fully. That's OK; you can still learn from those parts of a paper that you can comprehend.

10. Relax and enjoy. Perhaps this is the hardest advice to follow, especially when you're confronted with a complicated paper. But try to approach an article like a puzzle. It's going to take time and effort to make progress, but there's real satisfaction in doing so.

Resources for Students and Educators

WEBSITES

Research articles can be found at the following sites:
　BioMed Central (www.biomedcentral.com).
　Highwire Press (www.highwire.org).
　National Center for Biotechnology Information (www.ncbi.nih.gov).

BOOKS

The following books focus on the scientific method:

Carey, S. S. (2004). *A beginner's guide to scientific method.* Belmont, CA: Wadsworth/Thomson.

Gauch Jr., H. G. (2003). *Scientific method in practice.* Cambridge: Cambridge University Press.

Giere, R. N. (1991). *Understanding scientific reasoning.* Fort Worth: Holt, Rinehart, and Winston.

Kitcher, P. (1993). *The advancement of science: science without legend, objectivity without illusions.* New York: Oxford University Press.

Kuhn, T. S. (1970). *The structure of scientific revolutions. International Encyclopedia of Unified Science*, volume 2, number 2. Chicago: University of Chicago Press.

Popper, K. (1959). *The logic of scientific discovery.* New York: Harper and Row.

Wilson, E. B. (1952). *An introduction to scientific research.* New York: McGraw-Hill.

The following books focus on reading and writing:

Adler, M. J. & van Doren, C. (1972). *How to read a book.* New York: Simon and Schuster.

Allay, M. (1996). *The craft of scientific writing.* New York : Springer.

Day, R. (1998). *How to write and publish a scientific paper.* Phoenix, AZ: Oryx Press.

Graff, G., & Birkenstein, C. (2005). *They say/I say: The moves that matter in academic writing.* New York: W. W. Norton.

McMillan, V. E. (2001). *Writing papers in the biological sciences.* Boston, MA: Bedford Books.

Pechenik, J. A. (2004). *A short guide to writing about biology.* New York: Pearson/Longman.

The following books focus on experimental design and statistics:

Ambrose, H. W., and K. P. Ambrose. (2002). *Handbook of biological investigation.* Knoxville, TN: Hunter Textbooks.

Barnard, C., Gilbert, F., & McGregor, P. (1993). *Asking questions in biology: design, analysis, and presentation in practical work.* New York: Longman Scientific & Technical.

Gould, J. L. & Gould, G. F. (2002). *Biostats basics: A student handbook.* New York: W.H. Freeman and Company.

Heath, D. (1995). *An introduction to experimental design and statistics for biology.* London: UCL Press.

Quinn, G. P. & Keough, M. J. (2002). *Experimental design and data analysis for biologists.* New York: Cambridge University Press.

Sokal, R. R. (2005). *Biometry.* New York: W.H. Freeman and Company.

Zar, J. H. (2006). *Biostatistical analysis.* Upper Saddle River, NJ: Prentice Hall.

ARTICLES ABOUT READING PAPERS

Gillen, C. M. (2006). Criticism and interpretation: Teaching the persuasive aspects of research articles. *CBE Life Science Education 5,* 34–38.

Levine, E. (2001). Reading your way to scientific literacy. *Journal of College Science Teaching 31,* 122–125.

Muench, S. B. (2000). Choosing primary literature in biology to achieve specific educational goals. *Journal of College Science Teaching, 29,* 255–260.

Mulnix, A. (2003). Investigations of protein structure and function using the scientific literature: An assignment for an undergraduate cell physiology course. *Cell Biology Education, 2,* 248–255.

Pall, M. L. (2000). The value of scientific peer-reviewed literature in a general education science course. *The American Biology Teacher, 62,* 256–258.

Russell, J. S., Martin, L., Curtin, D., Penhale, S. and Trueblood, N. A. (2004). Nonscience majors gain valuable insight studying clinical trials literature: an evidence-based medicine library assignment. *Advances in Physiology Education 28,* 188–194.

Smith, G. R. (2001). Guided literature explorations. *Journal of College Science Teaching, 30,* 465–469.

Woodhull-McNeal, A. (1989). Teaching introductory science as inquiry. *College Teaching, 37,* 3–7.

EXAMPLE RESEARCH ARTICLES

Bricelj, V. M., Connell, L., Konoki, K., MacQuarrie, S. P., Scheuer, T., Catterall, W. A., et al. (2005). Sodium channel mutation leading to saxitoxin resistance in clams increases risk of PSP. *Nature, 434:* 763–767.

Carrier D. R., Deban, S. M., & Otterstrom, J., (2002). The face that sank the Essex: Potential function of the spermaceti organ in aggression. *Journal of Experimental Biology, 205,* 1755–1763.

Fleischmann, R. D., Alland, D., Eisen, J. A., Carpenter, L., White, O., Peterson, J., et al. (2002). Whole-genome comparison of *Mycobacterium tuberculosis* clinical and laboratory strains. *Journal of Bacteriology* 184:5479–5490.

Pai, R., Tarnawski, A. S., & Tran, T., (2004). Deoxycholic acid activates β-catenin signaling pathway and increases colon cell cancer growth and invasiveness. *Molecular Biology of the Cell* 15:2156–2163.

Rabbani, M. A., Maruyama, K., Abe, H., Khan, M. A, Katsura, K., Ito, Y., et al. (2003). Monitoring expression profiles of rice genes under cold, drought, and high-salinity stresses and abscisic acid application using cDNA microarray and RNA gel-blot analyses. *Plant Physiology* 133:1755–1767.

Rao, P. K., Kumar, R. M., Farkhondeh, M., Baskerville, S., & Lodish, H. F. (2006). Myogenic factors that regulate expression of muscle-specific microRNAs. *PNAS* 103:8721–8726.

Rocheleau, J. V., Remedi, M. S., Granada, B., Head, W. S., Koster, J. C., Nichols, C. G., et al. (2006). Critical role of gap junction coupled KATP channel activity for regulated insulin secretion. *PLoS Biology* Vol. 4, No. 2, DOI: 10.1371/journal.pbio.0040026.

Schweitzer, M. H., Wittmeyer, J. L., Horner, J. R., & Toporski, J. K. (2005). Soft-tissue vessels and cellular preservation in *Tyrannosaurus rex. Science* 25:1952–1955.

Tarnopolsky, M. A., Zawada, C., Richmond, L.B., Carter, S., Shearer, J., Graham, T., et al. (2001). Gender differences in carbohydrate loading are related to energy intake. *Journal of Applied Physiology* 91:225–230.

Weiss, S. L., Lee, E. A., & Diamond, J. (1998). Evolutionary matches of enzyme and transporter capacities to dietary substrate loads in the intestinal brush border. *PNAS* 95:2117–2121.

Wittert, G. A., Turnbull, H., Hope, P., Morley, J. E., & Horowitz, M. (2004). Leptin prevents obesity induced by a high-fat diet after diet-induced weight loss in the marsupial *S. crassicaudata. American Journal of Physiology Regulatory Integrative Comparative Physiology* 286:734–739.

Woolley, D. M. & Vernon, G. G. (2001). A study of helical and planar waves on sea urchin sperm flagella, with a theory of how they are generated. *Journal of Experimental Biology* 204:1333–1345.